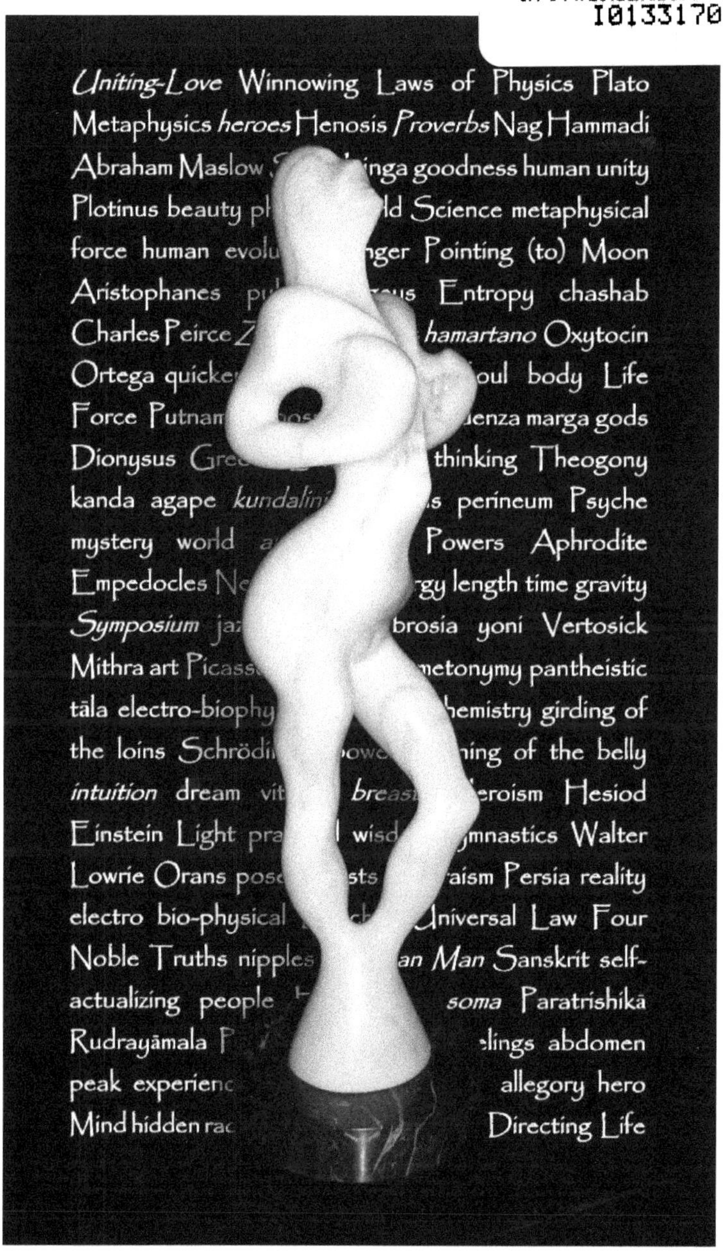

Uniting-Love Winnowing Laws of Physics Plato
Metaphysics heroes Henosis Proverbs Nag Hammadi
Abraham Maslow ... inga goodness human unity
Plotinus beauty ph ... ld Science metaphysical
force human evolu ... nger Pointing (to) Moon
Aristophanes pul ... us Entropy chashab
Charles Peirce Z ... hamartano Oxytocin
Ortega quicke ... oul body Life
Force Putnam ... enza marga gods
Dionysus Gree ... thinking Theogony
kanda agape kundalin ... s perineum Psyche
mystery world a ... Powers Aphrodite
Empedocles Ne ... gy length time gravity
Symposium ja ... brosia yoni Vertosick
Mithra art Picass ... metonymy pantheistic
tāla electro-biophy ... hemistry girding of
the loins Schrödi ... ... ning of the belly
intuition dream vit ... breas ... eroism Hesiod
Einstein Light pra ... wisd ... mnastics Walter
Lowrie Orans pose ... sts ... aism Persia reality
electro bio-physical ... Universal Law Four
Noble Truths nipples ... an Man Sanskrit self-
actualizing people ... soma Paratrishikā
Rudrayāmala P ... lings abdomen
peak experienc ... allegory hero
Mind hidden rac ... Directing Life

Directing Life

The hidden power of quickening, gnosis, union and love

by

Robert L. Peck
Leslie M. Cassinari
Christine S. Gavlick

Personal Development Center
Since 1975

Integrating Modern Science
with Recovered Ancient Science

Personal Development Center
Lebanon, Connecticut U.S.A. 2006
PersonalDevCenter.com

Publisher's Note

2 3 4 5 6 7 8 9 10

This second printing, March 2020, of the original 2006 publication of *Directing Life: The hidden power of quickening, gnosis, union and love* contains minor editorial corrections throughout the text as well as format improvements in the Indian Sanskrit translated document [i.e., *Haṭhayogapradīpikā*, *RigVeda* (*Ṛgveda*), and *Rudrayāmala* (*Parātriṃśikā*)].

In addition, the original frontispiece depicting the Indian *Shiva Linga* and Greek *Liknon* with enlarged text in the background has been replaced with a sculpture we feel depicts a human activating their inner *Shiva Linga*/Greek *Liknon* (located in the perineum) that must include breast and nipple stimulation.

Artwork and cover design by Leslie M. Cassinari.
Frontispiece sculpture by Richard Jaworowski.

Translations of the *Haṭhayogapradīpikā*, *RigVeda* (*Ṛgveda*), and *Rudrayāmala* (*Parātriṃśikā*) by Robert L. Peck

Personal Development Center
P.O. Box 93
South Windham, CT 06266-0093

ISBN 13: 978-0-917828-11-9 (Paperback)
ISBN 13: 978-0-917828-15-7 (e-Book)

Printed in the United States of America.

A Brief Description of the Authors

The authors: a research physical chemist; a classical musician, athlete and teacher; and a dancer, alternative-special education teacher met over ten years ago at the evening public classes of the Personal Development Center in which they were jointly committed to the researching and teaching of Eastern self-development techniques.

When they uncovered and translated a suppressed Indian document, it soon became clear that the basic physiological processes of the body had been deliberately mistranslated. The possibility that the West had undergone similar distortions of the ancient teachings of inner powers was suggested by the entrance of the West into the Dark Ages.

They then became interested in translating and studying the writings of early Greek philosophers and poets and found nearly identical underlying philosophies and explanations of the higher powers found within enlightened individuals. The discovery of the usage of the winnowing basket as an allegory in both cultures and the depiction of creative powers as a fleshy protrusion gave positive physical proof of their common views.

The authors' specialized knowledge of energy conversion, music and dance as well as the behavior and learning skills of children allowed them to piece together the various writings into a series of exciting and supportable insights into the creative and controlling powers of the self-actualized or enlightened individuals.

Table of Contents

Directing Life

The hidden power of quickening, gnosis, union and love

Chapter One
Inner Powers

There are a number of conspiracies being discussed today, yet the greatest and least discussed of all is how the creative inner power of individuals, including what it is and what it does, has been deliberately and intentionally suppressed. This suppression began, in part, when Religion decided that in order to save souls it had to first control the individuals possessing the souls. However, the first concerted effort by the early Catholic Church to obtain absolute control by military-type force led the world into the Dark Ages.[1] Over the centuries the methods and tools used for controlling individuals slowly evolved with similar types of failures but also with startling successes. These changes have led the vast majority of the members of the 'free' world into being under an absolute control that is far stronger than that first envisioned by the early Catholic Church.

Over twenty-three centuries ago, Plato warned the world of this loss of individual control as he wrote about a creative power within individuals and how rulers fear its emergence within their subjects.[2] He gave an exaggerated example of the creative power in his statement that a mere handful of heroes, when fighting at each other's side with that power, could overcome the world.[3] He gave a clue to the source of this power when he stated that the very word hero (Greek: *heros*[4]) points to the source.[5] These statements may very well have increased fears even in his time,

---

[1] Angus (1967) Ch. 6-7

[2] Plato *Symposium*, paragraph 64.

[3] Ibid., paragraph 59.

[4] Individuals with godlike qualities per Homer, Hesiod, Pindar etc.

[5] Plato *Cratylus*, section 398c

since only a few years later, Aristotle, in his *Metaphysics*, noted that theologians had started to deny the existence of a universal source of such a fundamental power.[6]

Plato and Aristotle were not alone in describing an inner creative power. Long before Plato, the book of Proverbs, the earliest writing used in the Judaic/ Christian religions, taught of powers within the bowels or heart that can shape and change an individual's life and future. Consider the following statements from Proverbs 23:7 and 16:9.

> "A man becomes what he opens (*shaar*) his heart to."
>
> "The heart directs (*chashab*) the path a man takes, but Yehovah[7] directs the steps."

It is certain that the later Hebrew heroes demonstrated the power in the open heart; however, none of the extant Hebrew writings state the placement and nature of the heart or any detailed description of this power.[8] Likewise, there is support that the Christians before the fifth century believed in similar powers. The best evidence lies in the *Gospel of Thomas* [9] that was believed completely destroyed by the rising Catholic Church but was secretly saved in the library recently found at Nag Hammadi. Plato's model of a handful of united soldiers overcoming the world is further exaggerated to make a critical point in the *Gospel* when it states that if two people can become as one or find union, they can tell

---

[6] Aristotle *Metaphysics*, Bk. 14, Ch. 4

[7] From Hebrew, *Ye*: the, *Hovah*: becoming, eternal, cause.

[8] Note that the Songs of Solomon (verses 10:1 - 22:16) only refer to Yehovah and never to Elohim or to Lord God.

[9] Robinson (1988) *Gospel of Thomas*

a mountain to move and it will.[10] Also, there are a number of verses citing inner powers, such as the statement that the powers you can bring forth from within yourself can save you. Another is that a good man brings forth goodness from his inner storehouse.[11]

To return to the modern world, it is unlikely that Plato and the other early writers could have foreseen the near total loss of individual freedom that exists in the modern highly controlled and mechanistic world. They, no doubt, were also completely unable to envision the marvelous and magical technological world that was wrought by the creative powers of a relatively few individuals over the centuries. The early writings around the world all agreed that the majority of people walked in darkness but that a few walked in the light. However, it may be that only in modern times is it possible to understand how blinding the darkness is and likewise how brilliant the creative light can become.

The inner powers of those few individuals who walked in the light were studied relatively recently by Abraham Maslow, the founder of Humanistic Psychology. He started his career by studying 'heroes' instead of the iniquitous or mentally ill as was, and is, popular in psychology. These exceptional individuals, who are able to 'open their own hearts,' he labels 'self-actualized' or 'fully human.' It is quite germane that the characteristics Maslow notes among these people are nearly identical with the early religious descriptions of the enlightened or virtuous[12] individuals found in most ancient cultures.

---

[10] Ibid., verses 48, 106. The becoming 'as one' was called *Eros* at that time.

[11] Ibid., verse 45. Compare Greek *tameion*: 'lower storehouse' in Matthew 6:6 (but translated as 'closet')

[12] From Latin *vir*, *virtus*: 'man, manliness, vitality' (hero)

It is also interesting to note that Maslow[13] considers creativity and the ability to obtain union with the world as the chief traits of the fully human, since these are also among the highest traits described by the early Greeks.

Maslow, for example, gives an explanation of how the fully human are able to create and do no wrong. He explains that an inventor is able to unite separate forces or objects together, that others cannot perceive, to create something entirely new to the world.[14] He further elaborates upon the powers of this process when he explains how inner-directed individuals are able to do no wrong as they manage to unite what they "want to do, what they need to do, what they enjoy now and will continue to enjoy" into a harmonious unity that lesser individuals are unable to do.[15] Maslow emphasizes the terms goodness[16] and unity[17] whose full meanings, he states, "are old to the philosophers but new to us."[18]

---

[13] See Goble (1970) Ch. 3, pp. 26-28.
[14] See Prigogine (1996) for a metaphysical explanation of the chaos out of which ideas are drawn.
[15] Goble (1970) p. 33.
[16] The ancient Indo-European root of good is *ghedh* (not related to god) which means 'to unite, join or fit together as one.' The word union includes the uniting in the future, while goodness only includes union in the present.
[17] From Latin *unus*, meaning 'one without separation or parts'
[18] Goble (1970) p. 24.

Chapter Two
Basic Good and Evil Powers

The ancient view of good and evil is so different from modern views that a direct comparison cannot be made. Instead, we will describe the ancient views with their logic and practicality before introducing the status of good and evil in the modern world. In order to do this, we must also discuss the terms beauty and union in a manner quite different from the way they are generally used today.

The old basic understanding of good, union and beauty can be perceived in the simple statement: "Good is anything which unites an individual or object with beauty." Union is a creative and intermediary force that allows elements of goodness to be manifested as beauty. Beauty pleases, stimulates, is timeless and serves as an opening to divine perfection. Goodness is the force behind evolution, creation and beauty. Without goodness there can be no value to life.[19] The elements of good and beauty are metaphysical absolutes[20] that only appear with the intention or dedication to obtain more from life. In the third century CE, Plotinus wrote extensively on the metaphysical nature of goodness and beauty using the terms as they were then accepted in the Graeco-Roman world.[21] He can be summarized as stating that beauty and goodness are the basis for an authentic existence that increases individual powers and the scope of life, while evil and ugliness are the opposite.[22]

---

[19] Plato, *Republic* Book VI
[20] Ibid.
[21] Angus (1967), Ch. II
[22] Plotinus *Enneas* 1.6

The steps of Proverbs 16:9 are directed by *Yehovah* and are certainly good. Goodness is like the rungs of a ladder that need to be used to reach the beauty envisioned with inner longing[23] or yearning.[24] Goodness is all of the changes that open the personal world and its scope in order that beauty can be obtained. The yearning releases inner powers that bring forth or create the elements of goodness that lead to the desired state of beauty. This can all be expressed with the quite common experience of an individual who has a deep desire to obtain some goal or enrichment of life and then reports the series of unexpected, good things that happened that provided steps to the goal. Goodness is often experienced as a result of someone else's actions, but more often it is the discovery of an inner power or skill that brings it about.

Goodness, beauty, union, evil, and ugliness were all considered to be metaphysical forces rather than something tangible that could be learned, taught or done. It needs to be remembered that metaphysical forces or powers are hidden, unseen, or non-physical yet still are able to be manifested in a physical form that can be perceived in the physical world.[25]

The above definitions can be further clarified by Science which furnishes examples that certainly match the ancient usage of the metaphysical terms. Let us, therefore turn to nature for an example of goodness and express it in terms that would have been quite acceptable to the ancients. An excellent example that everyone can identify with is the goodness of a tree mastering its environment and increasing its beauty.

---

[23] Meaning "to stretch the mind forward in time"

[24] A dedication including an emotional awareness of what is desired.

[25] See Plato's metaphysics.

Everyone has seen a tree which has one extended limb that was evidently seeking more sunlight. The growth of that limb is obviously good since it increases the energy that the tree can obtain. In order to explain that odd limb, let us anthropomorphize the tree in order to introduce metaphysical concepts which the ancients were quite adept at using.

Consider first that the tree is somehow aware of the need for more sunlight and somehow perceives a gap in the forest canopy and yearns to be able to capture the additional energy of the sun coming through that gap. The resulting yearning sets into motion a very complex metaphysical power that is called the Life Force in the modern world. The uniting power of that Life Force then somehow selects, brings together and manifests the goodness within the tree. This goodness consists of the proper forms of energy and the required building materials uniting at the proper site and at the proper time to construct the new growth.

Science is acutely aware of the tremendous evolutionary powers that are manifested with the growth and maturity of children. These powers are obviously controlled to a large extent by each child as well as to some degree by his or her outer world. Maslow noted that a critical level of externally supplied goodness must exist within each child's outer world for initial growth and maturing, but then a child must find an inner source of goodness if evolution to a 'fully human' state is to take place.[26] This individual growth of a child might be compared with the growth of a sapling which satisfies its basic needs in order to grow, but then some special inner forces of goodness are required if that sapling is to outgrow other saplings and become a fully developed tree in the woods.

---

[26] See Maslow's list of hierarchal needs in the Appendix.

The Nobel Laureate Erwin Schrödinger (a physicist) gives further definition and examples of the metaphysical powers within Life in his wonderful book, *What is Life?* His message can be related to the controlled growth of the tree when he comes to the conclusion that he, like the tree, is somehow able to direct the inner Life Force to control his body. However, it should be pointed out that the tree represents a more fundamental aspect of the Life Force that takes its energy directly from the sun which a human is unable to do. Schrödinger therefore defines Life (of the higher life form) as the creative and unifying power that must constantly overcome an opposing, destructive, unseen force known as Entropy. This is done by the continual consumption of other life forms to replace that which is constantly being destroyed.

This battle reflects ancient concepts of the interaction of the expansive inner power of Goodness and the contractive powers of Evil. This is particularly so since the scientific descriptions of increasing Entropy sound like the actions of some ancient evil god of destruction and chaos. For instance, increasing Entropy is blamed for rising disorder, separation and confusion in the world as well as the decrease in available energy. Scientists believe that Entropy will ultimately result in the decay of the entire universe into a homogenous mess or chaos.

There is some strange connection between goodness and evil or the Life Force and Entropy. Any living structure must have a force of assimilation as well as a force of elimination.[27] Similarly, evolution must have forces that engender the new as well as the forces that assuage the old. Perhaps one of the best examples of this is the difficult experience of the letting go of childish ways to become a full adult. There is a strong tendency to tightly cling

---

[27] Aristotle *Categories*, III.10

to those things that once seemed to enrich the world even though the world has changed.

Religions have universally described the inability to fully trust in the steps toward beauty as an evil force. The Graeco-Romans described this evil force as that which causes an athlete to miss the target when casting a spear. The original word in Greek is *hamartano*[28] which is translated today as sin. Thus, a sin was a force that caused the missing of a personally chosen goal. A further hint of the original view of sin can be obtained from the word *dikaios*[29] which was applied to an individual without sin. This word meant 'equal, even or well-balanced', as well as 'legally exact and precise.' Such an individual was called righteous, able to follow and stick to the right path leading to personal perfection.

Religions were also nearly unanimous in describing those individuals who 'missed the mark' as being in darkness, unable to see clearly the steps or the goal. Similarly, those who could manifest their inner visions or goals were declared to walk in light. However, the old religions, as well as Maslow, agree that those walking in the light are a very small minority who have nearly always assumed some responsibility in assisting others to find the light or the inner powers that provide the light.

There have been few philosophical writings concerning the powers that keep the majority of mankind in darkness. Historically these people have been viewed as being a homogeneous mass of humanity without individuality and labeled in the descriptive term, the masses. They seem to have been quietly ignored by philosophers perhaps because their state of having

---

[28] Greek, *άμαρτάν*

[29] Greek, *δίκαιος*

missed the mark was accepted as the responsibility of religion to define, explain and correct.

However, this silence was loudly shattered when a Spanish philosopher, Jose Ortega y Gasset, wrote a well-researched book[30] on the subject in 1929 CE. Ortega wrote about the changes that had taken place in the masses in Europe as well as in the U.S. during the preceding two centuries. His view of the social changes within the masses is unusual, since he views the changes from the position of a self-actualized or fully human, creative aristocrat. When Ortega therefore stresses how the aristocrats or the intelligentsia are losing power over the masses, his motives are certainly open to being questioned. However, he presents another startling observation that the masses have likewise lost power, and both the aristocracy and the masses are reacting more and more as a mindless mob. Both are driven primarily by 'Public Opinion.'

Further, those individuals acting as rulers of society are mostly specialists relying upon learned skills, wisdom and transferred power rather than being the *dikaios* (i.e., fully human individuals with developed inner powers who can find union with others and the world). Instead of leaders who 'see the big picture,' the modern world relies upon leaders who see only a very small piece of it and specialists who attempt to integrate the pieces.

According to Ortega, the loss of accepted leadership and individual powers is resulting in the agglomeration of people into larger and larger groups. We would describe it as people huddling together without a sense of leadership and in fear of the unknown. The larger the group, the safer it can appear, and so mass reactions are used to select beliefs, entertainment, lifestyles and

---

[30] *The Revolt of the Masses* (1993)

values. As individuals identify with groups or the masses, the less their inner powers are used and the less individualistic they become until they are in fact no different from others. Ortega also cites how individuals have become 'spoiled' with the effective largess of our industries and government that, in supplying them with the basic needs and security of life, have made them unappreciative of what they have. As with spoiled children, the masses also believe they have power because others appear to listen to their demands.

There has been a more recent related study on the masses[31] by Robert Putnam of Harvard which indicates a further rising problem associated with the electronic age. He describes the increasing membership of individuals into card carrying groups such as the *AARP* or the *Sierra Club* who have strong political power because of their membership. However, they have a non-interactive membership, which is increasing the isolation of individuals while also making them responsive to institutional publications and political stands. Putnam takes the name of his book from his observation that people are playing alone today instead of in groups and that interactive groups are becoming smaller. He gives an example that church membership is gradually declining with individuals defining their own religious beliefs and trusting other individuals less and less.

Ortega's and Putnam's observations are certainly applicable to today's teenage gangs. Modern teenagers are reacting in general as being spoiled with less and less concern for necessities and responsibilities for others. They can be compared to the card-carrying memberships of their parents as they follow the statements of entertainment or athletic leaders which they quickly adopt as their own. Perhaps the largest concern for teenage gangs

---

[31] *Bowling Alone* (2000)

is that they appear to be more motivated by the words of a rap star than by any member within their own gang (Rap stars are of course also highly dependent upon the opinions of their teenage fans).

It is doubtful if many people would disagree with our summary that in today's world most people live in their own isolated and powerless world. They are far more concerned by 'what someone might think' or public opinion than by any institutional decree or even their own thoughts or desires. It is also apparent that most people believe that they have everything they basically need and hence they have no need to consider any significant changes in their thinking, beliefs or lifestyles, particularly if effort is required. In their present world of darkness, there is certainly no concern for reaching for more beauty, goodness or finding their own inner powers. They judge themselves good because they heed public opinion.

Chapter Three
Concealing the Hidden Powers

Questions can be asked at this time as to why the metaphysical powers of evolution and goodness are not common knowledge today, and secondly how have they been hidden? The 'why' can be approached by contemplating Plato's warning that rulers did not want their subjects to possess or even know of individual inner powers and certainly not to find a union among themselves that exceeds loyalty to the rulers. Another reason for its hidden nature is that people are instilled with a fear of unknown powers, present in others, as well as within themselves.

The answer to how the metaphysical or unknown can remain hidden is that in our modern high technological society there is very little teaching of ubiquitous metaphysical powers. Neither Science nor Religion (currently perceived as the two major players in the nature of reality debate) accepts the existence of independent metaphysical forces. Thus, they are both quite similar when attempting to explain why or how things interact or happen. Religion of course speaks of the control by God's Will, but that is not much different than Science's control by the Laws of Physics. For example, a rock is heavy because of the Will of God or because of the Laws of Physics, yet neither is a true answer to a child's simple 'why?' Instead of really digging into such metaphysical questions, Religion attempts to define and modify God's Will, while Science attempts to understand and add to the Laws of Physics. Even Philosophy, which used to be primarily interested in exploring underlying metaphysical forces, appears to be now more interested in defining the physical nature of reality.

In order to unravel the metaphysical secrets of inner powers, the knowledge of modern physical Science needs to be utilized along with the ancient metaphysical philosophical writings. However, to do this we need to understand the philosophical foundation of science which is built upon the statement of the early Greek philosopher Empedocles that everything is built of basic metaphysical, unchanging and universal building blocks.

The building blocks were defined to be: earth, air, fire and water, which are the equivalent of today's 'dimensions' of mass, length, energy and time (which are also metaphysical quantities). Every physical thing and every physical interaction in Science must be defined in terms of dimensions or building blocks rather than causes and without concern where the basic building blocks came from or what they are made from.

Consider the following short and simple mathematical exercise used in defining gravity without any concern for what it is, where it came from or why it exists.

Science views gravity as the interaction of an object with the earth. A scientist then describes this interaction mathematically as **W=mg** where the weight is **W** and its rate of falling in a gravity field is **g** and its mass is **m**. This states the obvious to most people that the cause of physical weight is physical mass or that the more mass an object has, the heavier it is.

But what is mass? In defining mass, the circular inter-connectedness of Science is encountered, since mass can only be defined in its relationship to some other physical property. Mass is therefore described by the expression **m=F/a** which states that mass is equated to how much force, **F**, it takes to accelerate, **a**, the mass. This too is obvious as anyone discovers when pushing a car. Since weight is also force, and **g** is also acceleration, the two expressions amount to stating the same relationship, but without any description of what gravity, weight, force or mass consist of or might actually be. Science, in other words, can state that mass is the cause for the phenomenon of weight, but cannot explain how mass becomes weight. However, Science knows how they interact sufficiently well to send rockets to the moon.

This ignorance of Science as to the actual nature or 'why' of things is well hidden from the majority of people by the use of metonymies that seem to give an explanation for the hidden metaphysical effects. (A metonymy is another word that expresses the same meaning of a given word in a different manner.) Since the general public cannot understand the above mathematical expressions, Science explains that the weight of a rock is *caused* by its inner force of *gravity* which is a metonymy (Latin, *gravitas*, which means 'heavy'). Medicine offers simpler examples when it appears to explain that a muscle ache is *caused* by *myalgia* (Latin *myo algia*, meaning 'muscle ache') and baldness is caused by *alopecia* (Latin *alopecia*, 'hair loss').

Religions use a form of metonymy which is as old as the Greek philosophies. They basically explain the nature of things in much the same manner that the early Greek philosophers did by assuming the existence of elemental earth, air, fire, and water which God used to create things. A rock therefore is heavy because it is made mostly from the element of earth which is heavy. A muscle ache might be caused by the excess of the element fire which causes pain and baldness is caused by the excess of the element air which is thinness and the lack of solidity from the element earth. Modern religions of course cut corners and simply state that things are the way they are because God made them that way.

Chapter Four
The Ancient Science of *Techne*

Metonymies are an excellent introduction to a powerful, ancient Greek science, developed by early philosophers, that was used to explain and describe personal inner powers or other metaphysical forces. This science was so profound, and no doubt considered so magical that it was suppressed along with everything associated with the Arian Heresy or pagan powers by the early Catholic Church.[32]

The science, named *Techne*,[33] means 'skill, art, cunning or a method of making or doing something.' *Techne* is also a form of the root *tek-* as used in words such as *tekmerion* meaning 'proof' or *tekon* meaning 'giving birth,' which help to define *Techne*. *Techne* is concerned with how things come into being[34] and can be considered as explaining the source of power existing between cause and effect or how a hidden cause becomes its manifested effect.

Techne was a science of personal power that could be used to find creative answers or solutions to problems independent of the normal methods of learning. To the Greeks, *Techne* was a source of truth for the soul that had to be supported by or compatible with:

1) scientific knowledge,
2) practical wisdom,
3) philosophic wisdom and
4) intuitive reason.[35]

---

[32] Angus (1967) pp. 104-116.
[33] *τέχυή*
[34] Aristotle, *Nicomachean Ethics* VI.4
[35] Ibi

Techne was therefore not based solely upon imagination, wishful thinking, the supernatural or the occult, but rather upon a possible future.

It is important to note that a major aspect of *Techne* is similar to what is called transference in modern psychology but known as *Henosis*[36] in the Graeco-Roman period. Transference results from finding such a state of union with another person that thoughts, feeling and ideas could be as one, such as often reported in psychoanalysis.

The union or oneness required for *Techne* is, however, far more complex and includes the union with the self and world in the immediate future as well as the immediate moment as will be described shortly. The rise in the early interest in *Henosis* in Rome was credited to Mithraism[37] brought in by the military from Persia who had discovered that the effectiveness of hand-to-hand combat as well as personal survival was highly increased with *Henosis* as was taught in the early male-only Mithraic religion.

The basic concepts of *Techne* and transference also appeared in the philosophy of *Sol Invictus* or Henotheism,[38] as a further clarification of the powerful ancient philosophy and system. Constantine the Great gave the description of how metaphysical powers can be transmitted by using the sun as an allegory.

Briefly, the sun contains heat and light which is radiated across space as invisible rays to earth where they anoint or coat objects rendering them visible or warm. The important criteria in this model are that:

---

[36] Greek, *ενωσις*: (h)*enosis*, 'combination into one, union'
[37] Cumont (1956)
[38] Greek, *ενωσις*: (h)*enosis*, plus *theos*: 'God'

1) the heat and light in the sun is exactly the same as that which
2) is radiated and exactly the same as that which
3) lights or warms an object.

Henotheism, the religious view, considers that there is:

1) the nature or power of God,
2) the radiance of this nature, and
3) the covering or anointing[39] of an individual such that the same nature of God is known within.[40]

The *Techne* view is that:

1) an individual contains truth which is
2) manifested in some form which
3) can then stimulate or cover an observer with the same truth.

There are two separate aspects of *Techne*; one is its usage in creating and manifesting, while the other is the perception and understanding of that which was manifested by someone else. Both forms were considered to require metaphysical inner powers. In other words, godlike traits are necessary to use *Techne* to create and dispense the creation, while similar godlike traits are necessary to use *Techne* to perceive and decode the manifested.

An excellent introductory example of this is given in the book of *Genesis*[41] where the Pharaoh of Egypt had a vivid dream and

---

[39] Greek, *χριστος*, *christos*: 'to anoint or coat'

[40] See Peck (2004) for a detailed discussion of Constantine and *Henotheism.*

[41] Genesis 41 (King James version)

could describe it in detail, but he was unable to understand it. Joseph, on the other hand, was able to translate the dream and support it with "scientific knowledge, practical wisdom, philosophy and intuition" and hence could prove the truth of both usages of *Techne*. As a consequence, he was of course appointed to execute the vision of the Pharaoh. Both the early Greeks and the Hebrews considered that the source of the truth manifested by *Techne* was of some Divine power and would have had common agreement with the Divine-controlled steps of Proverbs 16:9.

Even though there is little public knowledge or support for the science of Techne, it is still being used today, such as by the self-actualized individuals, studied by Maslow, to manifest their inner goals. Creative scientists frequently use Techne and give evidence of it with their 'weird' explanations of how they attained their insights. Perhaps the best-known example is the story told by Isaac Newton when he experienced the falling of an apple as the same as the continual falling of the moon, which led to his profound theory of gravitation.

Many creations begin with what Einstein called 'mental experiments'[42] where the mind sorts through images appearing in the mind until one is somehow 'known' to work. This can be described as looking into the future to foresee which possible concepts will work or to find the solution that will exist with complete union with the problem.

Some poets are of course masters of *Techne*, such as Emily Dickinson who was able to impart strong metaphysical feelings with only a few words.[43] Many painters such as Pablo Picasso spent considerable time and effort developing their *Techne* to

---

[42] First defined by Ernst Mach as *Gedankenexperiment*

[43] See Appendix, Poetic Works of *Techne*.

depict deep inner metaphysical feelings and forces. Beethoven combined chorus and orchestra with the poetry of Schiller's *Ode to Joy* in order to manifest the metaphysical joy that unites one with the world which society has separated[44] in his famous Ninth Symphony.

The works of Newton, Dickinson, Picasso or Beethoven can only be fully appreciated, however, by a relative few who have the *Techne* to understand or find union with both the creation and creator. The initially created physical works of *Techne* exist as agents of transfer of the inner vision of the creator to the outer world, or as Aristotle noted, *Techne* deals with things coming into being. *Techne* serves both as an intermediary between a creator and the created as well as between the created and the observer. *Techne* is an intermediary agent that opens a path to the future or vice versa. We will shortly describe how *Techne* has been used in understanding supernormal powers.

We have avoided using the word 'Art' as is commonly done as a translation of the word *Techne*, since the majority of modern art is not *Techne* as originally described by the Greeks.

Walter Lowrie[45] gives an excellent dissertation on the usage of Art by the early Greeks and the subsequent changes that occurred when the metaphysical content of Art became replaced with a message,[46] teaching or something to be purchased. There can be

---

[44] "*Thy enchantments unite together what society's sword did divide.*"

[45] Lowie (1923) Ch. 1

[46] A modern example of this occurred when a clergyman in the late 1800's named Henry van Dyke arranged a new hymn as he took the music of Beethoven's Ninth Symphony *Ode to Joy* and replaced the words in the last movement as a way to provide a popular religious meaning instead of the original metaphysical experiencing of inner ecstasy and its power.

far more *Techne* in a child's drawing of a frowning face to express an inner feeling than in the regimented usage of art to attract, decorate or sell.

Because of the proliferation of art to educate or sell, the ability to use *Techne* to appreciate and understand the metaphysical content of art has nearly been lost. Art is now judged as to its physical portrayal and form rather than decoded as to its content. Paintings are judged according to their photographic likeness or how they will match the setting in some room. Writings, as will shortly be presented, are similarly judged according to how entertaining they are, how easy they are to read, and their political or religious correctness much more than their content.

The perception of art in terms of its physical form or popular acceptability is not new to this generation but must have been also present in ancient China, where the relative blindness of the public was aptly described by the maxim, "Finger Pointing (to) Moon." This maxim refers to the vast majority of people who can perceive only the physical form of *Techne*,[47] which is never the vision, but only that which points to the vision. The majority of people resort to describing and judging the physical finger, painting or writing according to social, religious or academic standards. Nevertheless, the maxim can be understood to be affirming the existence of fingers or allegories which are a product of *Techne* that do in fact point directly to the moon.

The metaphysical can only be pointed to and never be directly observed, touched or measured such as is common in the subjects of science. The science of *Techne* can be explained as the use of special allegories to manifest the metaphysical. If a proper

---

[47] Or as many state, "It takes too much effort to understand things." or "Why should I?"

allegory is created then it can be used to recreate the original metaphysical concept. For instance, we used an allegory above in describing force by stating that it is 'like' pushing a car. If you can relate to our allegory of pushing, then you can experience the same metaphysical force that we were experiencing.

An allegory often can present a clearer depiction of truth than can a lengthy technical essay, provided that it satisfies:

1) scientific knowledge,
2) practical wisdom,
3) philosophic wisdom and
4) intuitive reason as required by Aristotle.[48]

An excellent example of *Techne* is given in music with 'cadenza'[49] performed by a solo instrument or voice in a concerto (also found in modified form in early jazz and folk music). Cadenza comes from Latin, meaning "to fall or end," and has the accepted modern meaning of denoting the end of a musical expression specified by some arranger.

However, there is the earlier[50] concept that a cadenza was to be an unspecified, impromptu addition by the soloist to the *Techne* of the composer as well as to summarize and end the preceding expression. The soloist therefore must find an expression using *Techne* that is built upon the state of the musicians, conductor and audience to clarify or fully manifest the original metaphysical expression of the composer. The instrument or voice then

---

[48] See Aristotle, *Nicomachean Ethics* VI.4.
[49] Apel (1944) p. 120.
[50] Before the 19th century

becomes an allegorical tool for the further expression of the metaphysical content.[51]

A successful soloist must satisfy the above four requirements of *Techne* if an impromptu created cadenza is to fit into and amplify the concerto. The soloist, for instance, cannot ignore the physics or structure of the music, ignore the mood of the audience and past performance of the musicians, or be unaware of the metaphysical content of the music. The soloist must finally combine the preceding with creative insight and learned skills to fully manifest and amplify the vital metaphysical insight of the composer.

The self-actualized individual or the fully human person of Maslow can be described as a soloist within a society expressing cadenza to more fully manifest the steps or evolution of life. Similarly, any individual who initiates a change in the world with his or her creation is a soloist. A soloist, however, requires a supporting and united group that accepts individual powers and the resulting changes.

---

[51] Apel (1944) p. 742. The performer also varied the rhythm with *rubato*.

Chapter Five
The Inner Light

Although society offers security, material goods, training and education, these benefits are unable to stimulate inner powers or light within individuals. Geniuses and heroes who have appeared over the centuries were quite deprived of many modern societal benefits, although admittedly they were quick to use the necessary tools offered by society.

The best overview of the problem that must be faced by someone seeking inner powers is summarized with the *Four Noble Truths* of Gautama Siddhartha, named 'Buddha' or the 'Awakened' in the sixth century BCE.

The *Four Noble Truths* are literally translated as:

1) *dukha*: inner uneasiness or pain,
2) *samudaya*: accumulation,
3) *nirodha*:[52] no growth, and
4) *mārga*: a path or way.[53]

The first three truths are recognized by many people facing the midlife crisis. Nothing in their world seems to make sense, and they experience pain or disease as they contemplate it. The next awareness is that the pain and disease multiply and build up rather than being alleviated with their past procedures of avoiding pain. The midlife crisis reaches its maximum level when it is noted that they are no longer growing or evolving and rather seem to be slowly dying.

---

[52] *ni*: 'no,' *rodha*: 'growing, ascending, moving upwards'
[53] See also *Gospel of Thomas*, verse 2.

Siddhartha, however, offers a promise based upon his and many others' experiences that there is actually a method of continuing to grow and evolve. The later religion of Buddhism, however, became divided over what the method should be. The majority[54] took the easy route believing that the method for most people was to let the awareness of the outer world and self die, which of course eliminates *duhka* and *samudaya.*

The higher route of Buddhism[55] has taken several routes, but generally all accept the notion that an inner fluid, spirit or god must be obtained that can enliven or quicken the body and mind which is quite similar to the conclusions being reached by the early Greeks. Let us therefore review the thinking of both early India and Greece.

The early people honored and worshipped their rulers and heroes whose powers were explained by their existence as either gods or in between gods and mortals or at least having godlike characteristics. As mentioned above, Plato wrote that the Greek word for hero pointed to the source of the inner power of heroes, that is, the god called *Eros*,[56] as will shortly be discussed. Sanskrit likewise named a hero as a *śūra*, which was the name for the sun as well as a form of the word *sura* which meant god or goddess. The logic of hero worship should be obvious, since their deeds proved both the existence of inner powers as well as some unknown metaphysical source for their powers.

Both the ancient Eastern and Western cultures believed in a special fluid that when present in the body could cause evolution

---

[54] *Hīnayāna*: 'lesser method,' or in later Pali language, *Theravada*: 'way of the elders'

[55] *Mahayāna*: 'great method'

[56] The English letter *H* is commonly added in translation of Greek metaphysical terms.

of the body and mind to a godlike nature. Indians called the fluid *amrita*[57] which means immortal, and the Greeks called the fluid *ambrosia* which also means immortal. Other cultures spoke of an inner flowing energy, living waters or being filled with quickening.[58]

The presence of the transformational fluid was evidenced with the changes that took place in the body and mind when a heroic action was taken. For example, when an individual met some sudden emergency, threat or demand there was increased pulse rate, increased strength, increased awareness, a glimpse into a future outcome and changes in breathing. The assignment of these properties to an inner fluid was no doubt because of the observation of the overpowering changes in the body and mind that accompanied the drinking and absorption of alcohol. The power of alcohol was explained by the presence of a spirit in alcohol that could overpower an individual like the spirit of an indwelling god. Alcohol was therefore labeled as 'spirit', which is a label it carries even into the modern world.

An old trick for quickly obtaining heroic powers was to generate the metaphysical transformational fluid by direct means, which was commonly done by duplicating the physical reactions of the body found during heroic actions.[59] As an example, the practices of forced deep exhalations, churning of the belly or tightening of the loins (girding of the loins) were commonly used and still practiced today by many athletes and martial artists.

There were also practices that stimulated inner organs that would in turn generate the transformational fluids which resulted in a

---

[57] Also called *soma*

[58] Modern physiology explains the transformation of the body by the flow of released hormones.

[59] Peck (1988) Ch. 22

more continual inner production of the energizing or quickening waters or fluids.[60] These practices are almost without exception declared anathematic by modern religions which resulted in them being considered as disgusting, childish or hedonistic.[61] This censure, no doubt, resulted in the lack of public discussion of them such as found by Maslow who noted the unusual refusal of the self-actualizers to discuss themselves.[62] Today we often hear this opposing view expressed in statements such as, "Don't get your guts in an uproar!" or "Inhale deeply in the chest and relax!"

Another method for obtaining heroic powers was to imagine a god within the body that was able to inject the fluid (or power) into the body and mind. This was commonly described as taking on the nature of a god or uniting with that god. This concept is still evident in our modern language with terms such as enthusiasm[63] or jovial[64] which mean inspired by an inner god.

Children and actors are excellent practitioners of this method as they 'become' another person. People commonly attempt to take on the courage of someone they know, for instance, in walking at night in a threatening neighborhood or taking on the calmness of someone they admire when facing strife. There is also the sudden overwhelming change in the body that many times accompanies the facing of a personal trauma or need. In such cases it is quite common for people to speak of how they were overcome by some unknown force or presence that flowed into them to give them unexpected added strength, wisdom and foresight. This ability to 'take on' the abilities of someone else is certainly diminished

---

[60] Ibid., Ch. 23

[61] Peck (2004) Ch. 2-3

[62] Goble (1970) p. 23. Maslow lists these people as so guarded that only indirect studies were possible.

[63] From the Greek, *en*: 'inner' *theos*: 'god'

[64] From the god *Jove*

through the adherence to being an important, staid and proud individual. Such individuals might remark later that they could not believe that something bad could possibly happen to them, so they stood firm in their own 'right' and consequently were incapable of doing anything.

The Greeks called specific and lingering inner forces or presences *daimons* or demons (now considered to be evil or bad forces). These *daimons* were described as being indwelling gods or goddesses that had powers for inducing special talents or capabilities within individuals. This usage might be compared to individuals who put off doing some difficult task until they feel 'inspired' or filled with the powers that are needed to do the job. The methods used to hasten that inspiration can be further related to ancient techniques to activate some particular inner demon or god.

Over the centuries the teaching of heroic powers was stopped. Instead, the people were taught about evil demons, devils or *asuras*[65] and how individuals required an external protector god rather than personal indwelling gods. Descriptions of the source of the earlier inner metaphysical fluids were also changed from being inner physiological or metaphysical fluids to being outer symbolic ritualistic drinks or food that contained the spirit of an external protector god.

As an example of how *daimonic* powers are used, consider how the Indian system of *Yoga* controls them. If a student of *Yoga* is having problems with some inner disturbing force or power, the characteristics of that force are imagined to be completely removed and transferred to a god where the bad traits are even further magnified. That self-created god is assumed to be able to radiate the same disturbing power back into the *yogi* (or some-

---

[65] Sanskrit, *a*: 'not' *sura*: 'hero'

times to others). The god is then labeled as a particular *Mahaghora*[66] or a 'Great Trouble.' The next step is to assume that the *Mahaghora* resides within the shadows of the brain, ready to pounce. From that point on, the *Mahaghora* is observed and studied as to what brings it forth and how it behaves. By depersonalizing and anthropomorphizing the objectionable trait, it becomes possible to learn to control it or even to use it in a positive manner.

Modern people experience what can be described as *mahaghoras*, but rather than referring to them as being inner gods or devils, they have become known as neuroses or psychoses controllable with drugs. However, there are virtually no references to inner powers that can raise an individual toward becoming as a god. We are all taught of all the horrible things that our minds can do, but never about inner powers that can make us great. Certainly, the great pharmaceutical companies are not searching for some inner fluid that can turn an individual into a hero or a god. They are too busy supplying profitable drugs that quiet the *mahaghoras* (even at the expense of inner powers).

---

[66] *maha*: 'great' *ghora*: 'terrible' (god)

Chapter Six
The Invention of Gods

As a preamble to this section, let us affirm that *Techne* can be used to create allegories that can in turn be used to become the source of greater powers than originally possessed by the creator of an allegory. Consider, for starters, the child who created an allegorical imaginary playmate who is then capable of forming new allegories which are perfect for the development of that child. There are many professionals who rely upon their created professional role to guide them to the solutions of problems that they could not solve without the professional role. We have already mentioned taking on the characteristics of a god or hero to give us greater powers. Authors often speak of creating characters for their books who then guide and assist in the writing of the stories. Creative people often cite their amazement at the power of some created allegory that led to other allegories that then led to what they sought.

The very important consideration that must be given to allegories is that the allegory in all respects can become more real than its creator. This is, of course, quite common as for instance, with celebrities or people in authority who find their roles more real than their normal existence. The tragedy of this common situation is that generally an individual credits the power of the role as coming from the outer world rather than coming from a personally created allegory. In other words, they believe that their power is only reflected or given by some outer forces, presences or gods. Since they believe this, their private life can be viewed as disastrous because they cannot find a source of inner power for their 'off-stage' or 'unprofessional' roles.

Let us now turn back to the eighth century BCE to the writings of the Greek epic poet Hesiod who is similar to the child being controlled by her imaginary playmate or the author controlled by his fictional characters. He presents a document which is obviously *Techne* developing a powerful allegory[67] to describe the inner nature and powers of individuals. Unfortunately, his work, although heavily referenced by later philosophers, has become reduced to being considered as a pagan and pantheistic myth.[68] Even the name of his document, *Theogony*, which has the literal meaning of *Begotten of God*,[69] has been erroneously translated as the *Origin of the Gods* because of this later interpretation of his work.

Hesiod creates a creator of the universe whom he names *Zeus* after the earlier fabled god. His Zeus is initially without a physical body and form and is helpless to manifest his visions and desires. However, his *Zeus* then creates the gods in heaven who then can be perceived to have created Hesiod. Hence, his work is titled *Begotten of the Gods*.

Reading *Theogony* must start with an opening understanding that Hesiod is describing the result of his mental game that yielded an allegory that fits his inner questions as to his own origin and powers. He, for instance, gives us an understanding that he has searched back into his own mind to a state of beginning where he exists only as a powerless awareness in what he calls *khaos* (chaos), an infinite space of unformed matter. He then offers the result of his *Techne* in his starting description of *Zeus* who must be the reflection of his own inner mind and its self-awareness and drive for goodness or expansion.

Zeus can of course be called by many other names. Science, for example, calls *Zeus* Universal Law, which is contained within

---

[67] Greek, *allegoria* (ἀλληγορία): 'veiled language' *υπονοια*: 'the real meaning which lies at the bottom of a thing, deeper sense'
[68] Greek, *muthos* (μύθς): 'fiction' [opposite to (λόγος) 'logos or truth']
[69] Greek, *theos* (θεος): 'god' *gonos* (γονος): 'begotten of'

every particle or quantum of energy or the interaction of the two. For instance, even during the assumed Big Bang of the beginning of the physical universe, Science describes creative reactions that follow a Law present within each particle or force.

As Hesiod points out though, a *Zeus* or a Law is not enough to manifest change or to bring matter and life out of the emptiness of a creative mind or power. There must be some other power that can connect a present vision with its future manifestation or a cause with its final effect. To manifest the vision of the bodiless *Zeus* in chaos, Hesiod describes how *Zeus* had to create the inter-mediating god *Eros*. *Eros* can then be perceived as an allegorical anthropomorphizing of the energy and intelligence necessary to direct and manifest the visions of *Zeus*. In terms of the Big Bang theory, the laws that energies and particles must obey can be attributed to the creator of the Laws of Physics or *Zeus*, while the actual merging of energy and mass can be attributed to some power such as the 'dark energy of the universe' or some power such as an *Eros*.[70]

The ancient concept of *Eros* was revived with the work of Charles Peirce, an American physicist, philosopher and mathematician who in the late nineteenth century described a special kind of an evolutionary Love which he called *agape*.[71] *Agape* is not, however, the same as its current usage as charity or Christian love, nor is *agape* the same as the modern dictionary definition of *eros* meaning limited or sexual love. Peirce describes the Greek term *agape* as a self-determining creative or evolutionary love that serves as both the force for spontaneity as well as the cause for change. *Agape* is credited with being an agency for change or

---

[70] As an example, Einstein's $E= mc^2$ equates energy with physical matter and assumes a controlling intelligence and Law.
[71] Hausman (1993) pp. 171, 173-177.

evolution and is connected with the mind which he describes as the place for the "fountain of existence."[72]

[72] Ibid., p. 172.

Chapter Seven
The Creation of the Creative Individual

The brilliancy of Hesiod was not so much that he defined *Zeus* who in turn defined the nature of the gods, but that he was then able to turn religious definitions into philosophical axioms. He must have used a system of *Techne* that assumed that a religious statement was fundamentally true (even though he had just originated it) and then took it one more major step. That is, *Zeus* created Hesiod on earth. In accepting this as a truth, the next question would have to be that if this were so, then how could he, Hesiod, as an individual created by *Zeus*, be independent of *Zeus*, as to both consciousness and creative power? How could he as a product of *Zeus* be able to create *Zeus* and then prove the reality of *Zeus*?

This type of question is asked by many people as they, for instance, find self-fulfilling prophecies or learn the truth of Proverbs 16:9 that they can direct their own lives. The thirteenth century *Zohar* of the Jews likewise taught that there was a Divine nature and power within individuals and interpreted the story of the *Garden of Eden* as the genesis for humans rather than their fulfillment. The *Zohar*, for instance, quotes *Genesis* 3:5 that although Adam and Eve were created by gods they could also become as gods by first breaking free of the garden by eating the fruit of the tree of the 'knowledge of good and evil.' An experience of *Techne* provides the awareness of the ability of an individual to create a solution to a question that can then become real and then provide unbidden answers back to the creator. Or that someone else's creation can be reached through *Techne* such that it becomes a real teaching and force of its own.

Hesiod's next step of defining the individual is difficult for modern seekers to accept, because of the broad conditioning of our culture that there is no form of creativity existing outside of heaven. To offer a quick opposing view, it should be noted that a simple example of creative intelligence has already been given in the text of the intelligence and creativity of a plant to reach more sunlight. In his convincing book,[73] Frank T. Vertosick, Jr. offers data demonstrating that any life form must have intelligence, including cancer cells or bacteria. He summarizes his view that the world is in fact controlled by *intelligent design*, but that it originates within the separate intelligences of the life forms on earth.[74]

There is an inherent problem in ascribing inner powers to humans and that is that humans die. How can a human have properties of a god if he or she dies? The answer which was nearly universally adopted by the ancients was that humans had a soul that was free to leave the body at death and that the soul was a god or at least had godlike properties. The inner soul could then be credited with the creative powers of humans and the variations of powers among individuals could be explained as variations in inner souls similar to the differences in the mythical gods in heaven. This view was maintained through the early Christian era in Rome as evidenced in the catacomb drawings of the soul or psyche leaving the body at death, to be described later.

The later Catholic Church, however, declared that the soul was an integral part of the physical body and that there were no individual metaphysical powers other than those of the remote God. This view resulted in extensive changes in views of the individual as well as the world. Many of the current debates as to human nature are a reflection of this early decree. For instance, the early Catholic view in limiting an individual to his or her physical body removes even the possibility for any inner meta-

---

[73] Vertosick (2002)
[74] Ibid., p. 8.

physical powers such as the presence of a higher mind or awareness beyond the physical capability of the physical brain.

With the loss of metaphysics in the modern Western world, the acceptance of inner metaphysical powers was lost and certainly the acceptance of an independent soul that was a god. In terms of a soul, many people speak of having a soul which is of course opposite to the ancient beliefs where the soul took on or possessed a body. Popular science supports the concept that the physical brain is the center of intelligence and self-awareness and does not publicize evidence to the contrary. There should be little doubt in anyone's mind that our present society identifies individuals and their value with their physical bodies and brains.

However, let us continue with the assurance that at least some people manifest inner powers that are in fact godlike. The question then is what is the soul if it is not some function of the brain? Hesiod offers an excellent allegory that can be used to reach deeper and deeper into the meaning of life and what is beyond.

Hesiod had to use *Techne* to develop an allegory that could explain how an individual could have visions and awareness such as *Zeus*, be independent of *Zeus*, and yet draw inspiration and power from *Zeus* if required. The individual also had to have the power to physically manifest inner visions and then be able to physically interact with the creations. This leads to the heretical notion that the individual actually had to have more powers than did *Zeus*. (It must be remembered that *Zeus* was created by Hesiod, who then created Hesiod and then the allegory that Hesiod was greater than *Zeus*.)

His following description of the innermost nature of an individual is very difficult for most people to accept today and certainly for most men to accept. Hesiod makes the simple statement that the soul or center of all individuals (including brawny men) is a feminine psyche called *Aphrodite* with the attributes of the fabled goddess of perfect beauty.

He defines this personal *Aphrodite* as being created out of foam,[75] which is obviously an allegory about the interface between three differing worlds such as the foam on the shore of an ocean. Such foam consists of air, water and solids. *Aphrodite* as the psyche-/soul can then be explained as being created and existing between the metaphysical, mental and the physical world.

This allegory is readily supportable as we are aware of how our consciousness can switch from a dream to the marketplace or from the deepest of contemplations to howling with pain from a stubbed toe. It is also through this interface of our *Aphrodite* that we can sense the eternal chaos, our own eternity, as well as the source of basic knowledge of ourselves and an underlying sense of personal control. *Aphrodite* must therefore be considered as the triple intermediary between the body, *Zeus* and *Eros*.[76]

The creation of *Aphrodite* from foam caused by the churning of the ocean is also a common allegory for the generation of higher powers and, as will be discussed later, is based upon the finding of higher powers within individuals with inner churning[77] of the lower body.

This *Techne* is quite useful in explaining the difference between what religions call the enlightened and those in darkness. The

---

[75] First named *Aphro-genia* or 'foam-born' (Plato *Cratylus* 406c-d)
[76] *Aphrodite* can be compared with the Indian goddess *Lakṣmi*, also called *Rambhā*, the goddess of fortune and beauty born from the foam of the ocean. *Rambhā* is compounded from *ram* which means: 'to delight or enjoy carnally' and *bhā* which means: 'to manifest.'
[77] *Mantham* in Sanskrit. The chief *yoga* practice can be said to be *mantham*.

brain and physical body are easily overpowered by the outer world as religions acknowledge (and utilize). Under these conditions, *Aphrodite* becomes bored, restless or sleeps as the body and brain are reduced to being a robotic animal without the higher responses. On the other hand, if *Aphrodite* awakens and yearns for more, the body and mind unite with the nature of *Aphrodite* and her goal becoming voluptuous, ecstatic, quickened, and able to play to any game or to take on any role in life[78] utilizing the full power of *Eros*.

As the original concept of *Aphrodite* became distorted and lost, *Aphrodite* was placed in heaven as being a beautiful goddess but not as a source of beauty. The transformational power of beauty has been lost in the materialistic world and its sense of immediacy.[79] (*Aphrodite* also was renamed *Psyche* in Rome who then became half of the symbol of physical love as *Psyche* with the denigrated *Eros* as the god of erotic love as the other half.)

However, *Aphrodite* or the soul can by no means be fully lost since almost everyone knows that there is no end to beauty or that there is always more. As long as beauty can be recognized there must be at least potential beauty within the individual. This awareness of beauty which requires beauty to perceive it must be within our inner *Aphrodite* or the soul. Since *Aphrodite* is capable of seeking goodness she must therefore be basically beautiful herself so that she can recognize beauty.[80]

The ancient concept of a feminine soul was no doubt supported with the observation that masculinity was not a permanent or original state. A fetus for example, begins as a female, and mas-

---

[78] See Peck (2001) for a description of these characteristics.
[79] See Appendix, *The Scaffold for Perfection*.
[80] See Plotinus VI.2.

culine traits only appear with the production of testosterone and then later diminish with age.

The yearning of *Aphrodite* can be distinguished from desiring or lusting for something because of the separate feelings that accompany them. Yearning might be generally described as being feminine and the desire to unite with or become one with something else. Desiring and lust, on the other hand, is also generally felt as masculine with the drive to conquer or possess.

Aphrodite is like the original *Zeus*, bodiless and incapable of manifesting anything into the physical world. *Aphrodite* must therefore have an *Eros* similar to that of *Zeus*. Hesiod solves this problem by having *Aphrodite* give birth to her own *Eros* through the power of her yearning. This definition is no doubt quite acceptable to many individuals who have found that an intense and maintained yearning does in fact precede the attainment of major changes.

The superhuman or self-actualized individual described by Maslow must therefore have a soul that yearned sufficiently to have given birth to an inner *Eros*. This union, which is more commonly depicted as the union of *Psyche* and *Eros* rather than *Aphrodite* and *Eros*, distinguishes an evolving individual. The *Psyche* can set the direction to be taken in life and the inner god *Eros* can control the steps such that the direction can be maintained as described in Proverbs.

This basic interaction became the basis for the later coupling of *Eros* and *Psyche* (or *Aphrodite*) in art as the symbol for physical love. *Eros* was then, however, tragically degraded to being portrayed as a *Cupid* with no more intelligence or power than to shoot an arrow into someone to instill lust.

The model of the individual consisting of a feminine soul in union with a masculine god that unites the soul with the outer world was the generally accepted view for most of the world until the entrance of the Western world into the Dark ages. The new view was done simply by renaming and removing *Eros* or the various equivalent descriptions of *Eros* into a remote, heavenly, mediating and unifying god outside of the individual. The soul was also defined to be the same as the integral physical body and brain. This alteration in the Western world is well documented, occurring during the first five centuries of the Christian era with the rise of the Catholic Church[81] which included the burning and alteration of objectionable writings of its declared pagan adversaries.[82]

---

[81] Lowie (1923) p. 8.
[82] Angus, (1967) pp. 105, 383.

Chapter Eight
Union

Four centuries after Hesiod, Aristotle in his *Metaphysics*[83] credited Hesiod as being the first individual to consider goodness and unity or *Eros* as a funda-mental principle and quoted the lines of Hesiod's *Theogony*:

> "First of all things was khaos (*chaos*) made, and then fertile earth...and then *Eros*."

He then summarized the importance of Hesiod's teachings by stating:

> "Among existing things there must be from the first a cause which will move things and bring them together."

Plato extended Hesiod's model of *Eros* in his lengthy manuscript *Symposium*. *Eros* was carefully defined as being the symbol of a powerful and mystical force that was able to unite separate entities, whether in heaven or on earth, as one. *Eros* could therefore be considered as both a remote god as well as an inner personal god. The power associated with *Eros* could unite a lover with whom or whatever was loved and then both into a unified world that was longed for. Since *Eros* was considered to represent the power to create perfection by his ability to unify the necessary elements, *Eros* was depicted as being a hero with all of the qualities of perfection within himself. (That which the word hero[84] points to, as quoted before, is of course *Eros*.) *Eros* (or the

---

[83] See Aristotle *Metaphysics*. I.4; also, Plato's *Symposium*, paragraphs 54-55.

[84] The letter '*h*' is added to *eros* to denote a metaphysical nature.

inner creative power), if he is able to give perfection, must be described as containing all of the elements of perfection.[85] Plato had the perfect example of a few heroes defeating a much larger foe at Marathon in 490 BCE, where greatly outnumbered Athenians not only defeated the Persian army but did it with very few casualties.

Gods who are the equal to *Eros* are to be found in many of the ancient cultures, perhaps exemplified with *Mithra* in Persia or *Mitra* in India who were also perfect mediating gods known for their power of uniting. These two gods were also considered as having heavenly as well as inner personal powers.

The *Symposium* of Plato explains that there are two natures of *Eros*. One is constructive and good while the other is destructive and evil. The difference between the open and hardened heart in Proverbs serves to exemplify these two natures that can either evolve or destroy an individual. Plato gives an excellent statement of the destructive power of *Eros*.

> "*Eros* directed to self is the underlying cause of all the offences done by an individual; for the lover is blinded by the beloved, so that he judges wrongly…and thinks that he always ought to prefer himself to the truth."[86]

He also described this power of *Eros* as possibly becoming a personal tyrant living within the individual as a lawless lord and being himself a king, leads the individual on, as a tyrant leads a State. *Eros* therefore symbolizes the inner power that is capable of uniting the desires or wishes in the heart, good or bad, with the future and making them real.

---

[85] Plato, *Cratylus* section 398c
[86] *Republic*, Bk. 9, paragraph 60.

Historians argue that soldiers exemplified the desire for union or *henosis* with both their inner forces as well as with the outer forces of others and the gods. Soldiers had to rely upon and be at one with their own powers, those of their comrades as well as the powers of their leaders and rulers.[87] (It should be remembered that the rulers were generally perceived as radiating a divine power either as recipients of that power or of being divine themselves.)

The dominant Roman religion of Mithraism[88] arrived via the army from the East and was altered with its integration with other mystery religions which were similarly seeking union or *henosis*. The mystery religions recognized a unifying power in ritual often obtained during the sharing of meals and wine as well as in secret rites.

Perhaps one of the last constructive dissertations on *Eros* was by Plotinus, a third century Roman philosopher, whose writings became a main source for Christian mysticism.[89] Plotinus describes the nature of the soul or psyche and its relationship with Zeus or the source of creativity, as well as the ability of the psyche to create an inner *Eros* through seeking or yearning.[90]

There is one chief source of evidence of the early beliefs in union that survived the energetic burning and rewriting of documents by the early Catholic Church bent on erasing from the face of the earth what was called the great Arian Heresy.[91] It seems odd that the underground burial chambers of the Catacombs should have escaped the purge that swept the surface, but they obviously did.

---

[87] See Westbury-Jones (1939), *Introduction* and Angus (1967).

[88] Ulansey (1989 Ch. 2

[89] Angus (1967) pp. 32-34.

[90] Plotinus III.5

[91] Angus (1967) pp.105, 383.

In contrast to the modern isolation of individuals, the ancient Christian Catacombs[92] of Rome depict what was called at that time *henosis*, *agape*[93] or union.

The artwork of the Roman Catacombs depicts the feminine psyche as rising from the body. Psyche was never depicted in a sorrowful or humble posture but rather one of a joyous and expectant nature (the Orans pose) with the arms out flung and the head looking up and outwards.[94] This posture supports the reported feeling in the Catacombs of more joy than sorrow and produces an opposite feeling to the normal folded hands and bowed head posture of worship.

The artwork depicts liberation rather than the gloomy modern depictions of death.[95] For example, there is no depiction of suffering such as with crucifixion or a cross[96] and instead, there are a number of depictions of the Good Shepherd and Lamb[97] referring to the earlier depictions of the kindly ministrations and guidance of a personal intermediary God such as suggested in the 23rd Psalm of the Bible. In many of the drawings, the Shepherd is depicted as Orpheus whose followers taught of the separation of the psyche and body at death as well as eternal life for the deserving.

---

[92] Lowrie (1923) Introduction

[93] Retranslated later to mean "'unilateral, charitable, brotherly or Christian love."

[94] Lowie (1923) pp. 201-204. Plato in *Cratylus* argues that *anthropos* originally meant 'man looking up' from *anathron a opopen.*

[95] Lowie (1923) *Introduction*

[96] Parsons (1896) and Herbermann (1907) Symbols

[97] Lowie (1923) pp. 214-221.

Another common symbol in the Catacombs is the fish or pisces[98] which historically has symbolized union and metaphysical power,[99] the beginning of a new world, quickening, the spiritual realm as well as Christianity itself. Supposedly early Christians could identify themselves by scratching an arc in the sand and another Christian could acknowledge their universal union by drawing a second arc that intersected the first to form a fish in the intersection of the two arcs. (The intersection was known mathematically as the vesica pisces which has special geometrical relationships as well as symbolizes unity.)

The Catacombs contained paintings or drawings of small groups eating together that are known to have been depicting *henosis* or a special mystical union between the participants and 'Divine Love' or *agape*.[100] The Catacombs also contained small *agape* rooms dedicated to such feasts. These were quite similar to the earlier small Mithraic underground rooms used for finding *henosis* or union with others, as well as with the mediating god Mithra, through the ritualistic sharing of bread and wine.[101]

There is another piece of evidence of the early belief in *henosis* of the Christians that was discovered at the time of the writing of this book. A small inlaid tile floor was unearthed in the Megiddo prison grounds in Israel that was described as being the remains of a third century Christian Church by the Israeli Antiquities Authority. The center of the floor contains the images of two fishes or pisces that certainly identifies it with the early Christian movement. Also like the Catacombs, there is no indication of a cross, a controlling Priesthood or Jesus as God. Instead, the

---

[98] Ibid., pp. 232-236.

[99] Peck (2001) pp. 4, 58.

[100] The union was called *Eros* at that time by the Neo-Platonists and other groups. See Plotinus III.5.

[101] Westbury-Jones (1939) p. xxxvii. Also, Angus (1967) pp. 248-253.

inscription states that it is dedicated to the prophet (*chrestos*),[102] by a lover of God (*philotheos*). The dedication also mentions a table rather than an altar indicating its existence before the formation of the Catholic Church and its probable usage as a gathering site for seeking *henosis*.

The rejection of *Eros* and union in the West is perhaps easier to understand by considering a very similar rejection of the teaching of individual inner powers that took place in India. (It seems easier many times to view someone else's problems a bit more objectively than your own. India also did not burn or destroy their own records as did the West.)[103]

There was a very large movement in northern India dedicated to finding not only the union of the inner forces but the union with the world and heaven as well. This movement became known as *Tantra*[104] which developed several important insights into the various energies of the body, similar to the powers of *Eros*, and their control.

The science of *Tantra* is declared to be built upon the finding of the total union of everything in the consciousness of the all.[105] We will speak more about the uniting force and energy control of *Tantra* later as well as its relationship with the early Dionysian movement of the West. However, for the moment consider the documented institutional methods used to discredit *Tantra* which are characteristic to those methods used against other groups teaching of inner and unifying powers.

---

[102] Perhaps falsely construed as *Iēsous Christos* or Jesus Christ.

[103] See the Appendix for how the suppression was done.

[104] *tan*: 'to radiate' *tra*: 'to form together.' Also, scientific theory such as used to describe astronomical works.

[105] Sanskrit, *sarvaṃ sarvātmakam*

The chief tool that was (and still is) used to discredit unwanted groups or concepts is to label the teachings with a name that is socially unacceptable and their members as having licentious conduct. As an example, in our modern world, almost everyone would avoid anyone associated with *Eros*, Dionysus or Bacchus because of the images associated with them of: drunkenness, sexual orgies and even torture and murder.[106] *Tantra* publicly stated its goals of finding:

1) inner energy,
2) union,
3) ecstasy,
4) voluptuousness, and
5) the ability to take on a new identity with five Sanskrit words all beginning with the letter 'M.'[107]

The opponents of *Tantra* claimed, however, that the words actually meant the unforgivable Hindu sins of:

1) eating fish,
2) engaging in forbidden sex,
3) becoming intoxicated,
4) eating meat, and
5) taking aphrodisiacs.

There was no literal support for the alteration of the meaning of the five words and no support for the alleged crimes since such stories inevitably require that they took place in secret meetings.

---

[106] Nilsson (1957)

[107] 1) *matsya* (pisces, 'vital force'), 2) *maithuna* ('close union, oneness'), 3) *madya* ('ecstasy'), 4) *māṃsa* ('flesh' perhaps as *māṃsakandi* or *shivalinga-kuṇḍalini*) and 5) *mudrā* ('symbol of existence or authority').

There was a similar battle in fourth century Rome against the widespread acceptance of what was later called the great Arian Heresy or Arianism. This battle can be viewed as the major battle of the belief of the majority of the early Christians in the personal inner powers of Jesus as a man as opposed to the belief that Jesus was God. (An excellent, but obviously biased, overview of this battle is contained in the history of Athanasius,[108] who appears to almost single-handedly—and ruthlessly—fight for the present Trinitarian belief of the Catholics.)

The Indian terms of *yuj*, *yoga*, *yamala*, or *maithuna* which have the basic meaning of a power and union have suffered like *Eros* and *Henosis* as they have been redefined to refer to externally defined behavior and thought. For instance, *yoga* has become an externally imposed discipline as has the Christian *agape*. *Agape*, now called Christian love, is now generally defined as the giving without thought of recompense rather than being united with self and others. The word love has become what can be described as directed charity in various forms.

---

[108] Herbermann (1907), St. Athanasius

Chapter Nine
Food for the Soul

The phrase 'food for the soul' is still extant in our modern language, yet it, like the ancient terms for the drink of the gods such as *ambrosia*, nectar, *amrita* or *soma*, has come to mean something physical that is to be consumed rather than something metaphysical. The ancient usage of the term is exemplified by Plato's statement that food for the soul came from *Techne* or the *Arts* as compared with the food for the body which came from gymnastics.[109]

Aristotle describes the soul, *Psyche* or *Aphrodite* as a despot and separate from the intellect and body.[110] The concept of the separate soul and its feeding is quite evident in children as well as the feeding of the body. In general, a child today is controlled by adults through the limiting or altering of the food for the body such as by physical punishment, administering treats, restricting movement, expression etc. rather than by controlling the food for the soul. Children, on the other hand, learn to control one another with food for the soul or with *Techne*.

This control of each other starts with a question, such as, "Do you want to play and have some fun?" This question can be observed to awaken any sleeping *Aphrodite* with the yearning for 'fun and games.' As *Aphrodite* awakens, children become enthused[111] with the inner yearning of *Aphrodite* which then initiates the birth or energizing of the inner *Eros* or creative energy. It is this process that allows children to expand their understanding and

---

[109] *Republic*. See also *Colloquy of Manos and Una* in Poe (1983) p.704 which discusses the wisdom of Plato's statement.
[110] *Politics* I.5
[111] From Greek *en theos*: 'inner god'

experiences of their own future. In essence, games can become an accelerated form of the integration of the will of *Aphrodite* with the creative power of *Eros* under the tight control of the mind which dreams cannot do.

The process of playing games is mimicked in the adult world with the *Arts* such as drama, music, art and literature which are quite apparent in the child's world. The appearance of 'fun' is, however, normally forgotten with being a dignified adult. Without 'fun' the *Arts* as *Techne* degrade to the well-disciplined arts and their judgments. This results in the loss of *Techne* and its food for the soul. *Aphrodite* therefore sleeps, while the body and intellect assume control.

'Having fun' must be considered as the principal food for the soul, which is probably well known although seldom discussed. 'Fun' is a physical response that is suppressed in the adult world to maintain dignity. In the children's world, however, having fun or experiencing joy is a physical response, more so than an intellectual response. It is this response which must become the major source of food for *Aphrodite*. In the 18th century CE, Friedrich von Schiller, whose words to *Ode to Joy* were used in Beethoven's Ninth Symphony, wrote about the power of play and union that lead to beauty which is the one state that makes individuals whole.[112]

It is quite possible that the attempts to find a source of inner energy by imagining an inner god or by churning etc., as discussed earlier, might be bypassed by becoming joyous as a child playing a game and directly feeding the soul.

---

[112] Schiller (1967) Letters 14-16.

Approaching ancient societies with the view that joy is feeding the soul or finding God suddenly opens a much different view of ancient festivals as well as eating and drinking together. The ecstasy reported in such groups as the Dionysians, *Tantriks*, Arians or the followers of *Sol Invictus* can be viewed as the success of some early techniques for shedding the outer cares of the world and being able to play and evolve together.

Friedrich von Schiller considers play as essential to finding union with the self and the outer world. He states that in playing, the beauty of life is found which then guides the individual into being able to triumph over time and able to gain control over the objects in one's personal life. However, he warns that without food for the soul, which he calls 'quickening refreshment' (German: *Erquickung*), the sensitivity for beauty is lost.[113]

The techniques of rulers to remove the food for the souls are obvious by viewing the prohibitions of the early Catholic Church which managed to bring the world into the Dark Ages. The Church managed to starve the souls of the Western world by removing the tools necessary for adults to play. These tools are of course: public singing, dancing, drinking, touching and any other voluptuous or sensual behaviors. This also explains how Plato and his teachings became so unsavory since it was these same attributes that Plato and later Platonists advocated.

The Church then advocated purifying the soul for future rewards with suffering and condemning any worldly joy and ecstasy. This method of subjugating individuals to external control was utilized quite successfully in controlling slaves. The American slaves, for example, were forced to give up their native songs and dances and instead accept only church approved activities. They were of

---

[113] Ibid., Letters 25, 26

course also forbidden to drink or have any outward demeanor of sexuality.

Chapter Ten
Scientific Support

Before continuing, it seems wise to introduce some modern research that supports the power of *Eros*. In Plato's *Symposium* Aristophanes cites *Eros* as being a key power in the healing of ills, and indeed, this is generally accepted as true, since *Eros* is a product of yearning and dedication which are recognized as important to healing. Everyone is well aware of how death is either hastened or postponed by the determination of an individual. The bedside manner of physicians and nurses or the healing power of touch was long recognized as of equal if not of greater importance than medications.

A review of the official *Vital Statistics of the US* (or any other country) demonstrates the existence of a power to live that is independent of medicine. Once individuals have passed childhood and been able to direct their own path in life, the years that remained to them was about the same whether they lived in 1850 or in the modern world.[114] At the age of sixty, for instance, people in both groups were expected to live, on the average, about twenty years more. The very slight increase in expected years to live today is credited to lifestyle more than to medicine. Infants and children have always been highly subject to diseases, and their longer expected lives today is primarily the result of proper sanitation and not medicine.

Evidence of the inner power of individuals is also indicated in the required publishing of the clinical trial data on the efficacy of a new drug. The report must give the percentage of people who obtain the same effects of the drug (without side effects) when

---

[114] Most references to life expectancy ignore this fact and instead cite life expectancy from birth.

given a blank or inactive form of the drug given to others. This effect is generally always present and is called the Placebo Effect in which an individual finds the expected results ascribed to a drug, without having taken the drug.[115]

The existence of a power of yearning or the intention of an individual to change the self as well as the outer world was actually studied and verified recently at Princeton University.[116] Their results were so minimal when compared to the publicized but unproven claims of the popular psychics (such as levitating heavy objects and knowing future events etc.) that the news media showed no interest and hence the public is largely unaware of the brilliancy and fundamental importance of Princeton's work.

Briefly, hundreds of individuals were placed before a precision machine whose slightest variation in operation could be measured. The individual was then told to vary the operation of the machine (without touching it) such as typified with the intention to make a precision clock run faster or slower. Amazingly the clocks or devices could be made to change according to the intention of the individual. However, as typified with a clock, the rate of change was only on the order of a minute a day, which is certainly not of interest to a news reporter.

Nevertheless, there can be little doubt that Princeton demonstrated that an individual with a yearning or intention can open his or her heart to change the future world (with something like *Yehovah* or *Eros* controlling the steps). However, it should also be pointed out that spectacular results can be obtained even with a small rate of change if the intention is continued over the

---

[115] See *Persuasion & Healing*, an excellent book on the Placebo Effect as well as metaphysical union.
[116] See Jahn & Dunne (1989).

periods of time such as what is required for a person to obtain his or her heart's desire.

The Princeton work is also highly suggestive, since united couples working together could produce more than four times the variation of single experimenters. In addition, they also ran a number of experiments without machines that demonstrated the ability of separated individuals to mentally link their minds to a small degree on what one of them was seeing. This work also had the surprising result that this mental coupling could be independent of time.

The ability of the soul to find union with other souls still exists in our tightly controlled modern society and evidenced in the metaphysical phenomenon of 'first love.' In introducing this subject, an old, 1920's song called, "That's My Weakness Now"[117] comes to mind. Its lyrics describe the amazement of a man who has fallen in love and then starts discovering how the things that he didn't used to like have become his weakness or objects of desire since she likes those things. It is this type of change that occurs in lovers which might have provided Hesiod with his initial concept of *Eros*

'First love' can be described as starting with an attraction for each other that escalates into an overwhelming yearning. It is difficult to define what is yearned for other than being together. It is common to hear lovers describing what the yearning is not, rather than what it is. For instance, it is not sexual, not pleasure and not satisfaction. It is not liking, since it is much more. Lovers are very much aware, though, that they are experiencing some powerful force that is changing them.

---

[117] See Appendix.

One startling discovery of lovers is how their respective pasts can now fit together as supports for each other such as how past weaknesses can be seen as strengths. There is also such a strong certainty of this union of their pasts that both lovers can tell everything about their personal histories without fear. One essential requirement for first love at this stage is that the lovers are able to freely create a mutual goal and then manifest their own created future together.

Plato described this union with an allegory of the reuniting of the two halves of an original being who was split in two by *Zeus* to reduce their pride.[118] Just as reported by those experiencing their first love, each half fits or unites perfectly with the other. What is also difficult for many people to understand is that first love is the manifesting of the yearning for total union which is not based upon sexual intercourse.

Modern science is actually finding a biochemical basis for union that supports the old concept of a creative inner fluid. A number of experiments have been made on injecting the biochemical oxytocin into rats as well as studying its natural occurrence and effects in humans. The findings are quite metaphysical, since the rise of oxytocin results in the increase bonding of rats and humans particularly noticeable between a mother and child.

Oxytocin is produced naturally with body contact and physical stimulation such as evidenced by a mother and her infant(s). Proper contact and stimulation results in deep pleasure for all parties that is difficult to describe and to physically measure, although the bonding is easily observed. The mental and physical effects of the production of oxytocin are not yet evaluated other

---

[118] *Symposium* 192e-193a

than the calmness and increased awareness of the nursing mother, which seems natural for the protection of both.

A chief source of oxytocin is the breasts, which are also recognized by the ancients as the source or seat of feeling and thought.[119] Some of the reported methods of touching that increase oxytocin suggest that primitives and no doubt ancient groups were aware of them and deliberately used them to find ecstasy, union or *henosis*. Consider, for an extreme example, what would you have done in the evenings if you lived in a small hut without electricity or in a small village without controlled entertainment?

In our modern world, with physical contact highly discouraged and with constant judgment applied to what is said or expressed, it is not unreasonable to assume that many people find the most closeness with the physical sharing of food and/or drink and particularly with alcohol or social drugs.

Some might argue that social drugs and alcohol reduce social constraints. However, alcohol and drugs also reduce the ability to develop or maintain a group dedication to the future and this prevents total union.

---

[119] See Chapters Fourteen and Sixteen.

Chapter Eleven
Doubting Mind

Modern children are taught that they should think before acting or doing anything. Our whole society equates thinking and controlled actions with a superior nature. Perhaps everyone has encountered the famous statement of, "*Cogito ergo sum*," or "I think therefore I am," attributed to Rene Descartes that equates thinking with being alive. However, these words are taken out of context and in fact give a contrary meaning to his actual intention.

Descartes wrote to assist people to break free of false wisdom, and his *Meditations on First Philosophy* urges individuals to doubt[120] or to extend thought to include other possibilities or ideas. He argues that doubt can lead an individual to the truth of his or her own reality and existence. (Aristotle describes this type of mental functioning as 'speculation.'[121])

Doubting is not the same as criticizing, believing or finding fault, which assume only one truth. Doubting is not something that a normal conditioned brain is able to do, and institutions generally oppose all forms of it. Institutions are interested in enforcing adherence to their own viewpoints rather than creating independent thinkers. The institutional opposition to independent thinking might be illustrated by considering your horror if your own computer or vehicle suddenly questioned your decisions. Similarly, parents are horrified when their children begin to doubt their decisions. The act of doubting generally starts with asking a

---

[120] From Latin, *dubius*: 'hesitating between two alternatives'
[121] *On the Soul*, III.4

question[122] such as what is behind a thought or action, instead of the prompt acceptance of the original thought or response.

Plato used the word *erotao*[123] for this type of questioning which he suggested was related to a characteristic of heroes (*eros*).[124] *Erotao* or the seeking of union through questioning is seeking for cause or the metaphysical forces that brought about some condition or manifesting. This process results in the separation and understanding of what was manifested from what caused it to happen. *Erotao* is commonly utilized in research with a question such as, "What is really going on here?" In the ancient Eastern world this was called separating the sun and moon or the physical manifestation from the metaphysical cause or source.[125]

Charles Peirce considered that there had to be a discipline associated with finding answers through doubting, which is similar to the requirements relating to *Techne*. His requirements can therefore be expressed similarly as requiring: an intuitive and continuous direction of inquiry, philosophic and scientific wisdom, and practical knowledge that the answer must fit the real world.[126]

---

[122] From Latin, *quarestia*: 'to seek'
[123] Meaning to ask a searching, personal question.
[124] Plato *Cratylus*, section 398c
[125] See Peck (2004) Ch. 12.
[126] Hausman (1993) pp. 27-34.

Chapter Twelve
Creative Mind

Once the manifest reality has been questioned and its source or cause sought, there is many times a time delay before the answer arrives. Initially, however, there is quite often the awareness of some activity in the gut and the assurance that the answer is there somewhere. This is particularly so if the normal mental brain tricks have been tried and seem to lead in the wrong direction.

As a very simple example consider a case of deciphering the *Techne* of some political cartoon. You might struggle to understand it for some time, during which you review recent political events etc., to no avail. You finally seem to know that you can find the meaning, that the meaning can be found, but you must be patient. You then deliberately drop the searching and then later in the shower, you break out in laughter as the full impact of hidden *Techne* hits you and becomes manifested. At this time there is normally the awareness that it appeared from some place deep within your body, and not the brain.

Where is the source of the answers or understanding and what is it? There is surprisingly a large agreement that the source is the heart, however, not the heart in the chest but a central heart low in the abdomen. For instance, Plato places the center of creativity within or around the liver which he also considered to be the center for divination as well as the location of the soul.[127]

The word heart means center, and the center of the body was strongly diagrammed by da Vinci with his *Vitruvian Man* with legs and arms outstretched and touching a circle. Da Vinci draws

---

[127] Plato *Timaeus section* 71a – 72c

the center as existing in the lower abdomen, yet if the *Vitruvian Man* had been standing on his toes instead of being flat footed, the geometrical center would have been at the sacral level of the spine.

The sacral region of the spine is named for the sacrum bone and called the *ieros osteon* in Greek or the bone containing divine or metaphysical powers. Somewhere in history the soul must have been considered to reside within the sacrum or sacred bone, which makes excellent sense, since the interior of the bone is loaded with gray matter and is central to the major nerves of the lower abdomen.[128]

The *Ch'an*[129] system of China describes a cavity in the lower abdomen called the lower *tan t'ien*[130] which contains a cauldron that produces a creative and transformational fluid called *chi*. The *Rig Veda*[131] compares the source to an inner granary[132] which is where winnowing and churning take place producing *soma* and its storage as well as being the dwelling place of *Indra*. Epictetus, a first century Stoic philosopher, writes in his *Discourses* of entering a darkened room and locking the doors to find *Zeus* and the inner God. This is similar to the Christian expression in

---

[128] See Peck (2006), Chapter Eleven.
[129] Known as Zen in Japan
[130] Meaning field of the drug or the source of the immortal fluid.
[131] Bk. 1, Ch. 28
[132] Sanskrit, *ulukhala*: *ul*: 'burning,' *u*: 'protective force,' *khala*: 'granary floor'

Matthew[133] of entering the hidden treasure room[134] and locking[135] the door to pray.[136]

The allegory of entering a special room and locking the door is certainly expressive of how the desire for some answer must be separated from the functioning of the brain. The results can of course be increased if the room is a granary or storehouse containing that which is being sought for. The process of finding the answer in the inner granary was described as winnowing by both the Dionysians and the *Tantriks* as will be discussed later.

The book of *Timaeus* by Plato makes a clear distinction between inspirations such as are found in the lower room or area around the liver and the thoughts arising from the brain. He states that thinking interferes with inspiration, but also notes that the brain is essential in deciphering the products of the lower room or center once they appear.

The existence of a creative mind outside of a brain is evidenced by the 'near death experiences'[137] now accepted as furnishing evidence that there is awareness within an individual even if the brain is nonfunctional. In his book, *The Genius Within*, Vertosick mentions the intelligence and creativity found in brainless life forms such as in lowly bacteria. There is also the common experience of a source of insight that cannot be related to memory contained within the brain. The concept that the sacrum serves

---

[133] Matthew 6:6 (normally translated, however, as 'room' or 'closet')
[134] Greek, *tamieion*, τάμιειον: 'storehouse or treasury,' normally hidden and below ground.
[135] Greek, *kleisas*, *κλεισας*: 'to lock or secure a door'
[136] Greek, *prosenkhē*, *προσευχη*: *pros*, *προσ*: 'in the presence of' *enkhē*, *ευχη*: 'to beseech, take an oath'
[137] Ring (1982)

also as a center for intelligence and creativity should not be ignored.

Chapter Thirteen
Eros, the Hidden Source of Creation

Once the concept that creativity might exist in the lower abdomen is accepted, the next obvious consideration is whether there is a common source or power behind mental creativity and the sexual creation of life. There was a general agreement in the ancient world that mental creation and procreation came from the same creative source called *Eros*. Most people considered that the power of *Eros* would either go into procreative sexual energy or into mental and physical creativity; it was one or the other.

For instance, it was widely believed that the creative energy could be dissipated in sexual orgasms or that mental creation could diminish sexual response. This belief was no doubt the reason for the enforcement of sexual abstention of priests and the belief that holy thoughts would diminish sexual desire. Religious art, on the other hand, demonstrates that sexual and abhorrent scenes can be used to increase specific religious feelings and fervor which are certainly felt within the lower abdomen.

Maslow suggests some sexual correlation of creativity in his descriptions of what he calls 'peak experiences' during which individuals are filled with awe, ecstasy and are functioning fully in a unified world.[138]

Therapists also note how such experiences could remove neurotic symptoms and produce permanent improvements in their patients.[139] Maslow, however, notes that people cannot fully discuss such experiences but he fails to explain why, other than it is

---

[138] Goble (1970) pp. 54-55.
[139] Ibid., p. 56.

embarrassing. The ancients, however, could explain the source of the peak experience as rising from the sexual region which would certainly be embarrassing for a modern individual to describe.

The relationship between sexual stimulation and peak experiences can be supported by studies of the REM[140] dream state. During REM sleep, dreams become 'more real than real' and many times are described as a peak experience. These dreams have been found essential for mental health and many times are quite prophetic. However, what is very exciting in explaining the source of peak experiences is that even though there is no sexual drive or orgasm, the sexual region in both sexes becomes quite swollen and stimulated.

Before referring to ancient descriptions of the lower center in the sexual region that activates the peak, religious or creative states, we will discuss the activities of children, martial artists and performers that can help define the hidden creative center.

Children learn to stimulate themselves into becoming more responsive to the outer world. This stimulation can be observed through their exuberance, often rising with increased interpersonal body contacts, holding their breath and using forced deep exhalations, screaming, shaking, bouncing, spinning, jumping as well as falling on their buttocks. The techniques used by martial artists utilize enhanced but certainly more refined versions of the children's methods to prepare the body for competition. For instance, instead of bouncing on their fannies, martial artists sit cross-legged on a pad or skin that would provide pressure against the perineum and use various motions and breaths that would assist in what is called churning (or winnowing) of the abdomen.

---

[140] Random Eye Movement

Performing artists also stimulate themselves as they 'psych themselves up' or 'turn on' before going in front of an audience. Although no one seems to be able to accurately describe this process, there are a few occurrences that most would agree with. The first common experience is the feeling of sinking in the tummy leading to bladder pressure, even though the bladder may be empty or nearly empty. Because of the concern for loss of control, the perineal muscles are tightened, and the enjoyment of that tension is countered with fear of leakage of urine which further excites the whole lower abdominal region.

As the inner tensions of the abdomen change, there is the increasing strange feeling of butterflies in the abdomen or a vague sense of something rising in the gut commonly sensed as some form of nausea. To these sensations is added the 'warming up' as the part to be played or the role is practiced, which brings into play the breath and sensuality of the body. Although these symptoms sound very disturbing, they nonetheless can become looked upon as essential for preparation for the finding of the union with the outer world and the ability to fully respond to its demands and needs.

The urge to urinate and the above symptoms of performers facing an audience are also commonly reported by people suddenly facing some emergency or demand. The common experience of 'wetting the pants' is easily explained by the further tightening of the upper muscle in the perineum[141] which functions, in part, as the urinary accelerant muscle. Tightening this muscle forces out any retained urine left in the urethra but does not open the bladder. Tightening this muscle is, however, only an initial part of the much larger muscle and neural stimulation process that prepares the body for meeting an unexpected demand.

---

[141] PC or pubococcygeus muscle

Unfortunately, many if not the majority of adults have lost the ability to control and use the perineal and lower abdominal muscles and hence are unable to do much to meet an emergency. Arnold Kegel, a physician, recently demonstrated that as most Americans age, their perineal muscles weaken with disuse leading, in part, to urinary incontinence.[142] He recommended perineal exercises that not only correct most of the problems but also were found to increase vitality. Another related weakness in aging is the loss of the ability to exhale deeply. Many older people have allowed their lower abdominal muscles to deteriorate to such an extent that they can no longer exhale deeply or use the breath to stimulate the perineum and lower abdomen.

The facing of an emergency or any requirement for the usage of the full capabilities of your body is described with the biblical phrase of 'girding your loins.' Most people seem to have an understanding that this phrase means to tighten something about the waist or lower body. However, there is also the meaning of tightening and controlling lower body muscles to 'steel' or prepare the self, although modern society does not discuss it.

We will return to the physical and mental stimulation of the inner powers later in Chapter Sixteen.

---

[142] Kegel (1951)

Chapter Fourteen
Heart, Mind, Soul and Breasts

For a person to become fully human, control of the body must be shifted from responding like a trained animal or a cog in a machine under the external control of society to responding to the much quieter control of an inner mind.

It is quite true that the modern world is far beyond the dreams of the wisest philosophers a few millennia ago. However, the ancients would be aghast at the price that modern individuals pay for their technology as they identify themselves as being like the mechanical machines and computers of their own creation.

Machines and computers are identified because of their external control. It takes an external control to direct and control a machine or computer. Modern individuals are in fact very much like machines or robots controlled by society and particularly by what someone else might think. Unfortunately, they have become like machines in that they are not even aware of being controlled and accept physical explanations for their human behavior.

Instead of describing inner control in terms of electro biophysical chemistry, however, Chapter Seven introduced the concept of a soul that provides a metaphysical model for inner control. Chapter Twelve likewise described an inner mind that is found in primitive life forms as well as humans and which is not a product of a brain.

To add to these descriptions of the soul and an inner controlling mind, we will now introduce another radical metaphysical power contained in the breasts of both males and females. The breasts are not only intermediary sense organs between the external and internal worlds, but also the seat of the inner mind or soul.

Any references to the breasts are, however, completely ignored in modern translations of ancient Greek and Sanskrit writings. In order to bypass this problem, we will give the definitions of the basic major terms associated with inner power and let the readers come to their own conclusions. Consider, therefore the following list which is taken from Sanskrit and Greek dictionaries.[143]

(We found that the clarity of old concepts can be greatly increased by comparing both Sanskrit and Greek definitions since their basic philosophies have so much in common.)

- ***stēthos*** (Gr: στηθος) breasts[144] of both sexes, as the seat[145] of the heart, the seat of feeling and thought, (as we use heart).

- ***kardia*** (Gr: *καρδία*) heart, the seat of feeling and passion, inclination, desire, purpose, mind.

- ***hṛdaya*** (Sk) heart, center or core, soul, mind (as the seat of mental operations), the seat of feelings and sensations.

- ***thumos*** (Gr: θυμος) soul, the principle of life, feeling and thought.

- ***atman*** (Sk) soul, the understanding, intellect, mind, the highest personal principle of life.

---

[143] Monier-Williams (1990), Liddell (1996)
[144] Front part of the thorax, *θωραξ*, divided into two mastoi, *μαστοι*.
[145] Greek, *edra*, *εδρα*: 'seat, abode' seat of a physiological process. Sanskrit, *sthāna*: 'region, cause'

The *Greek-English Lexicon* recognizes the difference in the ancient and modern usage of the word heart in its comment that the description of the breasts is "as we use heart." In other words, the modern impression of the powers of the inner heart is the same as those that used to be attributed to the breasts. (It should be noted that we know of no ancient writings before the fifth century that place any metaphysical powers in the interior of the chest or beating heart. There are countless descriptions of a powerful inner heart in the human body, but they are not referring to the blood-pumping heart in the chest.)

It is difficult to equate the breasts with being the seat of feeling or thought in the modern world when the breasts of both men and women are relatively undeveloped. It is amazing that even the response of the nipples to temperature is generally unknown. The nipples serve in part as a sensor for controlling the blood flow into the outer surface of the body.[146]

Another ignored response of the breasts is the relief that is found upon beating, pressing or rubbing them in times of grief, panic or intense desire. The breasts can be experienced, as the *Greek-English Lexicon* states, as being the "seat of feeling and thought" with the production of oxytocin as mentioned in Chapter Ten. Connected with this may be the deeper fuller breathing that also occurs with stimulation of the breasts.

The breasts are certainly activated in embracing (even if the modern world attempts to avoid direct pressure). The power within the breasts is more systemic than local, and if their activity is noted, it can be perceived as occurring before the more directed activities of the lower heart. Chapter Sixteen discusses in detail how the breasts

---

[146] See Peck (1998) p. 109.

can be used, but for now consider only that the breasts are the source of the experiences that were falsely attributed to the beating heart.

The important consideration in this chapter is that the breasts are sense organs to the outer world which stimulate inner responses of the body. This function will be elaborated upon in Chapter Sixteen. It is also quite important to recognize that the mind of the soul in the heart can direct the entire body and brain to fully respond to the outer world as well as to a chosen future.

Chapter Fifteen
The Suppression of Union and Love

The nearly complete disappearance of *Eros*, as an inner creative power, can be attributed to its deliberate suppression by the early Catholic Church. Its loss certainly led the world into the Dark Ages. Today the capitalized word *Eros* is defined to mean the old Greek God, and the lowercase word is defined as meaning the normal love as attraction and affection for something. Words based upon *Eros*, such as erotic and erogenous, mean deviant or lustful sex, while *Eros* as a god has been reduced to Cupid as an initiator of lust. But the original meaning of an inner power as defined by Hesiod, Plato, Aristotle etc. is not described even in mainline English and Greek dictionaries.

Further, most translations of early Greek writings substitute the word love for *Eros* which compounds the problem of understanding the old documents and increases the chances of *Eros* being accepted as meaning sexual lust. The Greek writings that describe *Eros* must be read in the original Greek, or at least with the original meanings of key words, if the meaning is to be made clear.

It is even more tragic that the relationships that *Eros* describes have nearly disappeared along with the word. The closest evidence of its original meaning appears to be the term 'Platonic Love.' This term might have originally been useful to those individuals who had read the original manuscripts in Greek and could make the distinction between the modern term of love and the original unity and power of *Eros*. However, Platonic Love is generally used today to imply a close relationship without sexual lust, which is only a further limitation of the meaning of love and not at all descriptive of *Eros* which could include sexual relations. This shift in the meaning of ancient terms is quite common with modern translations that attempt to skew the original meanings to other desired meanings.

The universality of this technique can be illustrated by considering how early English translators of the Indian *Rig Veda* could not accept the concept of the inner fluids of *amrita* or *soma* and instead added and changed words to falsely teach that the source of such a fluid came from a psychedelic plant no longer available. This can easily be compared with the shifting of the description of the inner power of *Eros* to being the physical power of sexual lust.

The distorted translations of both cultures are supported by religious or academic claims of the infallibility of their translations and viewpoints such as are universally common in religious, political or academic translations or interpretations. The distorted translations in both cases (and others) are, however, quite obvious since they have almost no support from literal translations as well as none from scientific knowledge, practical wisdom, philosophic wisdom or intuitive reason. We have lost the old requirement of knowing Greek to be considered a true intellectual in order to overcome biases and distorted translations. Similarly, few high school graduates today acquire a basic comprehension of science and logic.

In short, the existing problem of the loss of the meaning and the experiencing of *Eros* is that the modern world has few teachings or examples of the creative union. Our society has replaced heroes with Hollywood pop stars. The news media finds eager readers when they denigrate some great person. Children are taught of the fallibility of past and present leaders. Perhaps only athletics remains somewhat free of the denial of the inner and quite exceptional powers of its stars, at least on the playing field.

A generally unrecognized example that exists today, much as it did in the time of Hesiod, is the union or *Eros* found within military life and which has no civilian label. This can be exemplified with what are now mostly historical battlefield stories of soldiers who stated how they 'loved' their comrades and would die for them.

However, soldiers are very reluctant to discuss such relationships today because of the dismissal of such relationships as homosexual or deviant (which they are not). Plato is accused of being homosexual in that he uses such examples of the power of soldiers filled with *Eros* who are able to conquer the world. Unfortunately, this example of Plato's is completely distorted today by those who argue that Plato is teaching that the power of *Eros* comes from the power of homosexual love. Such a statement can only be made by someone who has never experienced the inner creative power of *Eros* and cannot conceive of any power higher than sexual lust.

The old adage of the pen being more powerful than the sword is certainly proven by the ancient pen that replaced *Eros*, which unites and awakens the inner creative powers of the righteous, with the impotent word Love, which blinds and cripples the majority of the world.

Maslow bypassed this problem by noting the unusual relationships that self-actualizing people have with others but not labeling it as *Eros*. Instead, he stated that their close sense of kinship with others was of a nature that most people cannot understand, and which does not follow the normal social rules. He also noted that they are not highly motivated by sexual appearance.[147] Maslow also noted that the relationships of the super human individuals last far longer than those based upon

---

[147] Goble (1970) p. 31.

normal love and are immune to public cries to behave like civilized individuals.

Personal love today is primarily based upon a religious definition that the highest love is giving and caring for another person without thought of recompense or return. This definition is of course an excellent advice for the giving of charity and maintaining social structure, but it is certainly not a method of finding a union with a lover or an inner personal god.

Chapter Sixteen
The Ancient Non-Secret Secrets

We mentioned earlier that an allegory many times can convey far more information than a lengthy text, but it is also true that an object or action can often do the same without any dependency upon language or culture. As for example, the observation of someone crying manifests deep feelings within the observer that are generally the same inner feelings as those of the one crying such that the observer can then shed tears of sympathy or union.

Inert objects can likewise be agents of transferring deep inner feelings if the object is perceived by the observer to be the reproduction of a dominant element that was associated with the same earlier feelings. As simple examples, consider the changes that take place in the body upon observing food or sexual objects, either real or symbolic.

The ancient Indians had the Sanskrit word *linga* to describe this type of a metaphysical transfer from an object that resulted in physical, emotional or mental changes in an individual. The word *linga* has the common meaning as being a sign or mark; however, it also has a basic philosophical meaning as given in the *Sanskrit-English Dictionary* as: "the invariable mark which proves the existence of anything in an object (as in the proposition 'there is fire because there is smoke'), smoke is the *linga*." A sexual object such as an erect penis can therefore be a *linga* if it proves the existence of an inner sexual fire. The proof must of course be within the observer.

If an erect penis can indicate the inner existence of sexual power, is there a personal *linga* that can equally indicate the existence of higher creative powers within an individual? The question appears to be strange in our modern world since we rely upon external symbols or signs and not *lingas*. We trust our physician because he has a diploma as the sign of his power, or we know someone is holy because of vestments or an altar. We no longer make individual judgments of others and so we no longer look for *lingas*. This is quite fortunate, is it not, for those in authority, since they don't have to prove their capabilities?

There are, however, inner creative powers as well as *lingas* of their existence. As a starting illustration, most performers or athletes are very much concerned that they are properly warmed up or ready to perform. Perhaps most rely upon a warm-up technique or procedures, but many also rely upon a general sense of knowing or feeling that they are ready, which suggests some form of a *linga* or physical change in the body.

From our studies, the general feeling of readiness is a whole body and mind response, but there are also many who report a change in the lower bowels that feels like a rising flow of energy that says, "Go!" But what is the center or *linga* that indicates the preparation of the body and mind?

In the modern Western world, we have significant difficulties even beginning to answer this type of question because our culture has little vocabulary and few models to use for inner energy, much less what its *linga* might be. Further, everyone is so accustomed to accepting that the brain runs the show, that any notion to the contrary is generally met with scorn. Fortunately, because of modern research into the higher capabilities of individuals such as by Maslow, the revitalization of the perineum by Kegel, and the introduction of metaphysics into physics by quantum mechanics, the old writings are becoming more acceptable.

Let us, therefore continue beyond the discussion of the pubococcygeus or PC muscle that Kegel discovered to be largely undeveloped in his patients suffering with incontinence. In addition to the PC muscles, we must describe another perineal muscle of even more importance but with even less development. This muscle is called the bulbospongiosus or BS, and it lies between the PC muscle and the surface of the perineum.[148] The name of the muscle describes quite well the nature of the muscle. When the BS is activated, it is like a swollen bulb lying between the anus and the penis or clitoris.[149] When it is developed, the BS can swell to many times its original size and is quite soft like a sponge. With development and usage, it can become more sensitive and pleasurable to touch than the clitoris or penis. Women may initially confuse the swelling of the BS with the swelling of the labia, and men generally assume that this swelling behind the penis is the base of the penis (even though the BS can be swollen without penile engorgement). Since the BS is sponge-like, the nature and location of the swelling can be changed by opposing pressure such that the swelling can be localized and then thrust outward away from the perineum with the use of the PC muscle and other lower abdominal muscles.

Discussion of swelling of the perineum is largely limited to the fringe field of sexology which has documented reports of swelling during sexual stimulation. Sexologists, for instance, now report swelling so extreme in some sexually stimulated women that they are unable to be easily penetrated. The swelling in men is often falsely ascribed to the swelling of the base of the penis rather than to the BS muscle above it. Studies of the swelling of the perineum during the demand for inner power would of course be nearly impossible to pursue.

Ancient Indian sages did not have the Christian labeling of the sexual organs as sinful and were able to study and discuss

---

[148] In women the BS has an opening for the vagina.

[149] *Haṭhapradīpikā*, III, 109 gives the size as 9"x 3" and covered.

subjects such as swelling of the perineum with dispassion. Centuries later, the Christian aversion to this Indian viewpoint was certainly manifested in the Christian academicians' translations of the Indian documents as well as the development of Sanskrit-to-English dictionaries.

As an example, consider the *Sanskrit-English Dictionary* definition of *kanda*, which is the Sanskrit name for perineal swelling that appears frequently in the philosophical writings of India. The dictionary states that *kanda* is, "an affection of the female organ, considered as a fleshy excrescence, but apparently *prolapsus uteri*." (The assumption of a fallen uterus came of course from the male Christian author.)

In addition, other documents define the word *kanda* as the source of the powers that flow up into the body through three tubes (*nāḍīs*) symbolized in the medical caduceus.[150]

It becomes even more confusing to modern Westerners when the self-actualized, fully human or enlightened men are described as having a *yoni*[151] as well as a *kanda* in the original Sanskrit documents. This confusion arises because the term *yoni* has been retranslated to be a female pudendum, and its original meaning as a source or origin has been lost along with the knowledge of the ability of the swollen *kanda* to stimulate a higher power in the form of the secreted *soma*.[152] Both sexes are, however, actually quite capable of inducing swelling of the perineal *kanda* for the generating and evidencing of higher powers.

---

[150] See Woodroffe (1974) Chapters I and VII.
[151] *yoni*: 'female sexual organ, source, origin'
[152] See Appendix.

The swelling of the *kanda* is no doubt related to the swelling of the lips, the labia as well as the physically stimulated swelling in the breasts, which are classified as adenomas or swelling related to glandular secretion.[153] It is certainly possible that in all of these cases, the glandular secretion is that which the ancients called *soma*.

That the Indians may also have had confusion with the perineal swelling is suggested by the many different names that refer to it, such as *Shiva Linga*,[154] *svayambhu linga*,[155] *Agni jihva*,[156] *kundalini*,[157] and *jal gula*,[158] all of which are used to indicate the source or presence of a higher power within the body.

The term *jal gula* was, no doubt, the first term applied to the induced swelling as described in *Book I* of the *Rig Veda*.[159] This ancient text describes the power of *Indra* in the form of *soma* rising up from the *jal gula*, or covered bulb, after the lower abdominal muscles are churned. (*Indra* was the name given to the inner personal god that later became known as *Shiva*.) The more than one hundred verses of *Book IX* are devoted to further descriptions of the production and usage of *soma*. They repeatedly assert that *soma* is produced by being 'squeezed' or 'pounded' out of the swollen perineum, which sounds similar to how a modern physiologist might explain the production of a hormone from an organ.

---

[153] Lapedes, (1978)
[154] Sign of *Shiva's* powers
[155] *svāyambhū*: 'self-made'
[156] Tongue of *Agni*, the projection of the power of *Agni* (God).
[157] *kuṇḍalini*: 'the power of the coiled serpent'
[158] *jal gula*: 'covered bulb'
[159] 1:28:1-6. See Appendix for full verses and translation.

It was exciting to us to find that the Indians were not the only people to use the bulbous projection as a *linga* of inner power. The first step of this awareness started as we considered the common *Shiva Linga* Icon used in Hindu/Shiva temples. This Icon consists of a stylized, horizontal female vulva with a fleshy protuberance projecting out of the vulva. The common interpretation of the projection is that it is a phallus, but the shape is much more like a bulb.

This led us to consider another bulb of the early Dionysians of Greece that was also falsely claimed to be a phallus. This bulb was depicted rising up from the bottom of a wicker winnowing basket called the *Liknon*. The *Liknon*, filled with this cloth-covered bulb, was paraded and used as a symbol for the inner power found by the Dionysians. We quickly recognized the pelvic-shaped basket and the projecting bulb as another depiction of the perineum with the swollen flesh and '*yoni*' similar to the *Shiva Linga* icon.

There is another interesting symbol of the rising bulb from the vulva found in the hand sign called the 'fig hand'[160] which consists of the thumb protruding between the clenched fingers of the hand. The fig represented the female organ and the thumb the fleshy protuberance. This hand sign was used as a power that could ward off the 'evil eye.'

It is now known that the early people of the Indus Valley migrated out of the valley into the West and East, and the concept of the *linga* no doubt traveled with them, but surprisingly, so did the mystical allegory of the winnowing basket. The winnowing basket is missing in the English translations of the *Rig Veda*, but with just a bit of searching of the original Sanskrit it is quite easy

---

[160] Italian: *mano fico* or *mano figa*

to find references to it. In fact, *Book IX* of the *Rig Veda* is entitled *Soma Pavamana*,[161] which means 'Understanding Winnowing of *Soma*.' *Book IX* also has numerous references to the winnowing basket[162] as a tool for increasing inner powers.

The Greek name for the Dionysian goal, *orgia*, offers further confirmation of the concept of the swollen *linga*, since the term *orgia* is derived from the same root as used in words denoting: fertile land, organized, swollen and ready to produce, to be ready, basic propensities, ecstatic and mystic, as well as orgasm. This can be compared with the *Tantriks*' stated goal of *vama marga*,[163] or *vama chara*[164] which means finding that which is agreeable and good.

The word *orgia* became known as orgy with the modern meaning of a group being immoral and wanton. In addition to being accused of engaging in orgies, the Dionysians were also said to engage in eating raw meat, murder and torture. To that list was added another accusation that made us wonder about its source, which was that Dionysian women nursed wild animals (instead of their own infants).[165]

These accusations were nearly identical with those based upon distortions of the *Tantrik* term *vama marga*, which was claimed by detractors to mean taking the evil left-hand path consisting of orgies, eating meat etc. as already mentioned in Chapter Eight. It is no doubt because of the ease in falsely branding someone as evil that the ancient, enlightened individuals listed one of the

---

[161] *Pavamāna*: *pava*: 'to clean from chaff' *māna*: 'to understand'
[162] *sya* or *śūrpa*
[163] *vāma mārga*: *vāma*: 'anything dear, desirably good' *mārga*: 'a way, manner or method'
[164] *chara*: 'moving toward'
[165] Nilsson (1957) Ch. II

worse sins to be false witness (Greek: *pseudomai*) or false knowledge (Sanskrit: *avidya*). (It should be acknowledged though, that through the centuries many small groups have indeed sought sexual and/or drug debauchery falsely under the names of *Tantra* or Bacchus, the later term for Dionysus.)

In addition to the *Tantriks* and Dionysians, there were other groups that can be assumed to have taught similar evolutionary methods such as the followers of Orpheus and Demeter. These groups were labeled as mystery cults which were characterized by believing that humans had an inner source of creative power as well as an independent inner soul that could be freed to find an eternal heaven.[166] They were all later defined as secret cults even though there is no evidence that they hid their beliefs or practices.

The reason for depicting them as secret can now be understood as history proves that the early Christians used many of the 'pagan' rituals and beliefs. However, the later Catholic Church could not admit to this and so they used three methods to hide their own pagan origins. The first was to destroy all earlier records, the second was to say that the mystery cults stole the rituals from the Christians, and the third was to deny that anyone knew anything about them.[167] Fortunately, the Sanskrit literature of the *Tantriks* has survived. Unfortunately, it has suffered severe distortion in its translations to English by the colonizing and proselytizing English Sanskrit scholars (see Indian translations in the Appendix).

In order to unravel and clarify the actual methods and beliefs of the ancient seekers for enlightenment let us continue to pursue

---

[166] Angus (1967) Ch. V

[167] Ibid., Ch. VI. See also See also Westbury-Jones (1939), *Introduction.*

the *linga* and the winnowing basket. It is helpful to remember that these two icons were based upon the awareness of an inner power and the responsibility of an individual to generate, purify and guide it. It should also be clearly understood that this inner power was strongly opposed by institutions that wanted to have their members to be only aware of the power and control of the institution over them.

We have mentioned the problem of understanding a *linga* in today's outer controlled world, but there is also a modern problem with understanding the past popularity of the winnowing basket. Every family that bought or raised grain had to have a winnowing basket to purify the grain before it could be consumed. Its usage was quite simple and quickly learned, even though it involved a number of different forces. To our knowledge, one of the most recent references to the winnowing basket was given in the first century BCE, by Virgil as he agreed with the ancient statements that the winnowing basket that purifies grain is an excellent symbol for the purifying of the soul.[168] If the basic mechanism of the winnowing basket is kept in mind, corollaries between tossing grain and chaff up into the wind to remove the chaff and the use of the breath to remove worries and desires can be found. For instance, a modern method of purification reminiscent of the winnowing basket is the simple action of taking a few relaxing breaths and 'letting go.'

The nature of the winnowing of *soma* must be compared with the generation of *soma* such as described in *Book I* of the *Rig Veda* where it is produced by squeezing the thighs and moving the hips forwards and backwards.[169] This motion is commonly observed to some extent in some children as they sit and rock. Grieving

---

[168] Nilsson (1957) Ch. III

[169] See *Rig Veda* I, 28:1-6 in Appendix.

children can then add winnowing to the rocking as they take tortured and forced breaths during 'sobbing.' Deep and powerful sobbing is often described as being 'heartfelt.' To the rocking and sobbing can be added the pressing, squeezing or even pounding of the chest. For those who can remember such sobbing, manipulating the chest can be explained as intensifying the depth or power of the sobs.

The combined action of the lower pelvis and chest is the churning[170] referred to in *Tantrik* writings and can become a powerful restorative or energizing force. Similar actions in the body, but perhaps with not so much force, can be observed in children readying themselves to play or the performer getting ready to go on stage as already mentioned. Many people facing a severe challenge experience this inner churning and describe it as having their guts in an uproar.

The cloth covering of the *linga* in the *Liknon* assists in connecting the *Liknon* to the winnowing basket, *sya*, of the *Rig Veda*, since the *sya* is also described as being covered. Its covering is called a *surpa puta*, which means a cloth that covers the privates.[171] The need to cover the *linga* certainly suggests the human loin or the perineum out of which the *linga* projects.

To find a detailed explanation of the *Liknon* and its power, it was necessary for us to prepare a literal translation of the readily available *Haṭhapradīpikā*[172] used as a bible by students of modern yoga. (It should be noted that most translations of the book are based upon 'tradition' rather than actual words. For more information consult the Appendix.)

---

[170] *mantham* in Sanskrit

[171] *śūrpa puṭa*

[172] Also known as the *Haṭhayogapradīpikā*

The *Haṭhapradīpikā* gives extensive descriptions of the stimulation of the *kanda* or *yoni* to produce the *soma* or inner fluid that fuels the inner powers. *Books I* and *II* describe the common postures and breathing exercises taught in most yoga classes which prepare the body for the more strenuous stimulation exercises. *Book III* offers general instructions in the stimulation of the inner organs and the purification of *soma*; however, this third book is usually rightfully ignored because of its destructive or misleading translations which lead to a lack of relevance to most students.

Book III, in Sanskrit, starts with describing the application of direct pressure to the *yoni*.[173] This area is then essentially massaged from a moving foot and the churning of the abdominal muscles above it[174] to prepare the *kanda* to swell and secrete *soma*, or as often called the *kundalini*. The final swelling requires churning, winnowing, concentration and upward pulling on the lower muscles.

Book III introduces the *Liknon* in verse 8 which has the literal translation of: "Persevere and endeavor to obtain the valuable secret concealed in the wicker basket (Sanskrit: *sya*): the feminine energy responsible for uniting, taken out of the *sya* (*Liknon*), in such a manner as follows …" This is, however, commonly translated as referring to a jewel-filled treasure chest and being discrete like a proper lady. (See Appendix.)

As stated before, lack of interest in *Book III* no doubt arises because of the destructive and misleading translations of some of its key verses which are based upon an allegory of a created lower

---

[173] Most *yoga* teaching were written for men although they are equally effective for women.
[174] See *Haṭhapradīpikā* in Appendix for the actual text.

tongue[175] that can move in a lower space or the 'mouth' of the *yoni*. The original Sanskrit verses serve to describe the initial feelings and stages of the protruding or swollen flesh from the *kanda*. However, these verses and other verses relating to this lower formation and activation are horribly mistranslated and rewritten without any apparent regard for the original text. For instance, modern English translations teach that the frenum of the tongue in the mouth in the head should be cut a small amount each day until the tongue is capable of being pushed up into the sinuses. (Surprisingly, every English version of the *Haṭhapradīpikā* we have found uses this same unsupportable translation, which suggests either plagiarism or some united effort to hide the truth. Interested readers are advised to consult the Appendix for the literal translation and definitions of the words.)

There are a great number of people who have religiously done the various practices of *Yoga* as currently taught without finding noticeable perineal swelling of a *kanda* or inner *linga*. For instance, many learn to pull the tummy in and out or to alternately tighten the two long abdominal recti muscles in the belief that they are churning (*mantham*). However, despite all of this effort they are unable to find any perineal effects. There are also others who have used other exercises or lifestyles which have developed the perineal muscles and the power of exhalation, yet who are unable to effectively churn, find swelling of the *kanda*, or find the state of enlightenment (what Maslow called self-actualization).

The problem is that the lower muscles involved in productive churning consist not only of the two major perineal muscles but also inner muscles that lie just above these two. Most people are not even aware of the existence of the muscles within the pelvic

---

[175] The tongue of fire of *Agni* or *jihva*.

area and certainly cannot voluntarily move them. These muscles, however, can be noticed during vomiting, prolonged difficult bowel movement, clearing of the throat, or after prolonged deep coughing. These muscles are also stimulated by the reactions to strong emotions such as laughter and crying as well as to trauma, etc. as discussed earlier. However, as the majority of people age, it is obvious that these muscles become less and less stimulated and used. What is required is what a scientist would call a catalyst that is able to start a reaction without being directly involved with it.

The nipples and/or the breasts are the catalysts that are able to start or increase the inner lower churning. This is similar to their usage by the sobbing child mentioned above, but a far better model for adults is that given by the postpartum mother. After birth, the lower muscles are quite flaccid and unresponsive until the mother nurses her child. As soon as one of the nipples is stimulated, a response of the lower muscles is noted.

With continuing stimulation, the resulting activity of the lower perineal muscles then facilitates the restoration of the uterus and assists in the total recovery from delivery. (The *La Leche League* can be consulted for the many benefits that follow nursing including such things as reduction in problems with diabetes and cancer.) It is these same lower muscles that are of great assistance in churning to stimulate the production of *soma* or the inner vitalizing fluid.

Our culture, however, views nipples with the same abhorrence as it does the genitalia and cannot even accept nursing as a social activity. Ironically, we worship the fullness of breasts in young women but cannot accept the sight of nipples. A woman is judged for sexual adequacy by the size of her breasts, but not by the development of her nipples. Sexologists fortunately have been recommending the stimulation of nipples in sexual foreplay (for both men and women) yet there is no evidence that it is often done. In fact, the extreme tenderness of the nipples that most women face during the early stages of nursing certainly indicates the lack of working and toughening of the nipples. Modern bras, in hiding the nipples from view, also shield the nipples from stimulation and development. Men note the corresponding tenderness of their undeveloped nipples in activities where the clothing rubs repeatedly against the nipples.

Massaging developed nipples produces amazing physical response of the lower abdominal muscles as they tend to rhythmically contract and relax much as found by postpartum nursing. When direct control is found over the responding muscles, then further development of that area is possible. As the lower muscles are developed, another change is noted as the muscle tone of the area increases and muscle mass is increased. Indeed, women doing these practices find that their lower abdomen takes on the shape found during early pregnancy with a similar firmness. It is certain that the old expression that power lies in the belly refers to this type of development rather than to the modern flabby 'beer belly.' It is also quite possible that the early Dionysians did teach about nipple stimulation and hence the stories of women nursing wild animals were a highly imaginative interpretation of using the nipples to feed a higher or uncontrollable power.

Maslow noted, with apparent surprise, the absence of any literature about the superior person in modern science, and we noted, perhaps with even more surprise, the total absence of any description of non-sexual nipple stimulation in any philosophical or scientific writings[176] in the last 2000 years, except for nursing. Any books on the stimulation of inner energy can be assumed to have been at the top of the list for book burning or editing during

---

[176] It must be noted that modern sex manuals are describing nipple stimulation only as a prelude to coitus.

the great suppression of pagan practices. It was only because of some anonymous *yogi* in India who heard about our search and forwarded a book to us that we managed to find our first documented description of the ancient usage of nipples in developing inner powers.

The book forwarded to us, the *Parātrīśikā Vivaraṇa*, contained an inner book of only 36 verses dispersed throughout the outer book. This inner book, as we found out later, was called the *Rudrayāmala*, and was largely ignored in India by belief that it contained obscene and forbidden practices. The *Parātrīśikā Vivaraṇa* contained the original Sanskrit and what we considered as assistance in the translation of the inner document.

As we worked through the translation of the verses of the *Rudrayāmala*, it became apparent that it was a technical paper giving the basic definitions of fundamental terms as well as the basic underlying philosophy associated with most of the ancient world's religions. This includes: the existence and nature of the heart in the lower abdomen, the inner fluids, such as *soma*, *ambrosia*, living water etc. which supply energy to the indwelling goddess (comparable to *Aphrodite*), an indwelling god (comparable to *Eros*), as well as the soul.

The content of the *Rudrayāmala* can be summarized as stating that the source of the higher power in the body is produced by the swollen *linga* of the *yoni* which is stimulated by the physical stimulation of the perineum, churning, and the downward stimulation from deep exhalation. The stimulated *linga* is then responsible for the release of *soma*, *amrita* or *ambrosia* that is then controlled by the directing power of the inner mind, similar to the model of *Eros* being directed by *Aphrodite*.

Later we found that the *Rig Veda* also provides support for the physical stimulation of the nipples and does so in one interesting verse which describes that the inner churning and winnowing forces can be controlled by ten maidens who are able to stimulate the body to generate inner thunderbolts to produce *soma*.[177] The ten maidens are clearly the ten fingers, but the referral to them as maidens is very suggestive to those who have experienced the uncontrolled inner churning and winnowing. The ten fingers, gentle as maidens yet firm and dedicated, are able to stimulate the inner organs, muscles and nerves to produce explosive forces that resonate through the entire body.

The Appendix lists several of the cited Sanskrit verses which we recommend that the reader carefully consider as well as the References. The offered translations are so different than those endorsed by large institutions that the reader should verify what is written for peace and certainty of mind. We have given the Sanskrit words which can be easily verified in a Sanskrit-to-English dictionary. We have not given the full detailed information on any of the above Indian writings, since they require a longer discussion than what can be offered here. The Appendix or References should, however, give some assistance until another book can be written.

---

177 *Rig Veda*, Book IX, 1:7-8

Chapter Seventeen
Afterword

In conclusion, we, the authors, would like to play the mental game of how would the past heroes describe the inner powers? Is there some phrase or model that almost all individuals who have experienced the manifesting of inner powers would agree upon? What comes as an answer to us is a slight change and addition to the *Proverb* verse that we quoted at the beginning.

"The heart directs the path an individual can take, but an inner Divine Force directs the steps."

To that we would add Emily Dickinson's modern warning of the chief opposition to staying on the path.

The Heroism we recite
Would be a normal thing-
Did not ourselves the Cubits warp[178]
For fear to be a King-[179]

We are also quite sure that most heroes and enlightened individuals would agree that when souls unite upon seeking the same path, there is no limit to the experiencing of union, ecstasy and evolution.[180] Or perhaps some would state that just as one is able

---

[178] cubits warp: lowering the standards of measuring (the self)
[179] See Appendix, Examples of Poetic *Techne*.
[180] See Peck (2004) and Peck (2001).

to create one's own world and future, one is always also able to integrate that world with others and choose to create a world even beyond.

Many people speak of walking through life 'hand in hand' but we are sure that the fully humans would agree that it should include being of one mind and heart as well as having a common dedication. Then as the Princeton study suggests, their power, at least, quadruples instead of merely doubling and the mutual goal becomes a certainty.

Appendix
Ancient Writings on Inner Powers

One very surprising discovery about ancient philosophical documents is that many of them are actually technical dissertations that describe inner powers and how to attain them. This seems logical since inner powers were of a fundamental importance because of the requirement of personal responsibility and control in about every activity of ancient life. These personal requirements are now, however, largely replaced with reliance upon technological devices, external energy sources as well as training and education.

When reading or translating Plato's view of inner personal controls and powers, how concerned can a individual be for accuracy of understanding if the entire subject is considered to be of no value to the modern world? Similarly, can Aristotle's view of the nature of the soul be taken seriously in view of the well-publicized opposing views by both the scientific and religious institutions?

Technical documents differ from other documents in several characteristics. Firstly, they must be considered as existing outside of the normal literary writings which convey experiences within a cultural milieu. Technical writings do not contain a set, setting and characters to convey the message or story. Technical writings should be understandable therefore within any milieu, ancient or modern. Technical writings must describe the terms used within the document or make definite references to their sources. These definitions are, however, many times allegorical particularly in the expounding of new concepts. The usage of allegories, although many times quite essential, can also be used to distort the content for readers without the same rudimentary knowledge that the writer expects.

One of the best examples of this is the ridicule that has arisen because of Newton relating gravity to the fall of an apple. The allegory that Newton was using was that the moon continues to fall toward earth as does a falling apple. If a person does not know of centrifugal and tangential forces, the allegory can be rendered that Newton had to be hit on the head with an apple to realize that gravity makes things fall. The person citing the story may also have that characteristic smirk of being smarter than Newton, because of knowing that apples fall at an early age. Similarly, many people have the same smirk as they hear about the teachings of Hesiod, Plato, Aristotle and many others because of their own limited primitive and scientific illiterate view of the powers of the universe.

Because of the self-contained definitions and lack of cultural influences, the language of technical documents does become nearly universal. As an example, most scientists are able to read technical documents written in other languages even though they cannot speak the language or even read the idiomatic primers of school children. A scientist (or a computer) is able to translate foreign technical articles by first mastering the basic syntax of the language and then using a technical dictionary. However, on the other side, no matter how versed one might be in the literature of a culture, the simplest of scientific relationships can be lost as technical words are assumed to take on a much different cultural meaning. As an example, what does the simple technical statement using common words, "A pressure gradient across a barrier results in a potential difference" mean?

Ancient documents add one more requirement to their understanding and that is that most of them were written in the same manner as with the Greek *Techne*[181] that uses a particular form of

---

[181] See Chapter Four.

allegory. For example, ancient statements such as girding the loins or churning the abdomen mean very little to modern individuals who have lost much of the control of lower muscles as well as the knowledge of how to gird or churn. Unfortunately, most people today read such statements, not as stating an inner process but rather as tightening the belly or moving it in or out rather than looking for inner activities that correspond to the terms.

For those readers who have not experienced the actual physical feeling of girding the loins (as if getting ready for battle or going on stage) or of churning, consider the feeling and forces involved in walking in deep mud or muck and use those feelings as a model of what to look for in the desired activities of the lower belly. If you are able to create such feelings within your own body then, and only then, are you able to understand the writings, since there is absolutely no difference between the message of the author and your experience as a reader.

Perhaps the chief problem in reading or translating ancient documents is that the modern world has generally lost or denied the metaphysical connection between cause and effect. Surprisingly, it is science that is largely to blame for this problem. For instance, an athlete knows that a great deal occurs between a ball being thrown and a thrown ball, an artist knows that there is a great deal of difference of an effect and the effort to cause that effect.

Science, however, tends to make cause the same as the effect with its physical laws such as a body in motion tends to stay in motion, which can be stated as the cause of a body moving is because the body is moving. Similarly, the cause of heat being radiated from a hot surface is that a hot surface radiates heat. This lack of separation of cause and effect is quickly evidenced by a child asking, Why? Many ancient scientific documents, however, attempted to explain 'why,' such as why did the writer have inner powers or why is a metal heavier than water? The ancient *Vedas* of India stated that the expression or understanding of truth, *vidyā*, required a number of attributes[182] including verified science; logic and metaphysics; practical arts; and spiritual truth, which correspond closely with the requirements for understanding Greek *Techne*:

1) scientific knowledge,
2) philosophic wisdom,
3) practical wisdom, and
4) intuitive reason.

182 Or *trayī*; 'consisting of': *ānvīksikī*: 'logic and metaphysics' *vārtta*: 'practical arts,' *ātmavīdyā*: 'spiritual truth'

Indian Literature
Introduction

India does not have the history of burning or destroying documents that do not meet some political or religious point of view as does the West. It is quite likely that this is so because the Indian sages of that time accepted a universal expression of a universal and perennial religion that accepted:

1) the existence of a creator,
2) the creator's manifesting powers, and
3) the transfer or radiation of those same powers within enlightened individuals.

Like many of the Greeks, they accepted their own form of a metaphysical *Techne* to express these powers and that the resulting allegorical names and descriptions were quite secondary to the basic concepts. The discussion under the *Rig Veda* lists the wide variation in terms that were used and accepted.

India's ancient documents were secured from alteration to some extent because of the original built-in rhyme and rhythm that facilitated the oral transmission of the teachings or stories. However, this form then obscured the content because like poetry, the words, word order, and even the sounds of the words had to be altered to render the work rhythmic or poetic. Sanskrit is made even harder to translate because of the common practice of exchanging of letters within a word as well as the combining of words together with the change in the ending and beginning letters of the combined words.[183]

---

[183] There are rules for uniting words called *sandhi*.

To further compound the problem, many times only the critical words are listed and seldom in the order of modern usage and without connectors, verbs or modifiers. The result is an ancient puzzle requiring the guessing of what the verse means and then testing to see if the words fit the guess. If not, another choice must be made. If the general meaning of the verse is unknown, then obviously no amount of guessing will help. But even with knowing the subject, it was not uncommon for the authors to spend over two hours per verse. The task was then only possible because of the usage of a computer-based Sanskrit-to-English dictionary[184] which would allow the searching for the basic roots of words.

The search for basic roots made it possible to overcome some of the errors introduced into the widely used *Monier-Williams Sanskrit Dictionary*. The nature of the problem is best introduced by quoting from Monier-Williams in his *Introduction* when he describes that his chief aim as an Oxford Professor was to translate "our sacred Scriptures" into Sanskrit to aid in the conversion of the natives of India to the Christian religion. As further confirmation of his aim, he states later that he did not deem that it was worthy to edit or translate the *Rudrayāmala* which we found to contain definitions of key words quite in opposition to his.

Monier-Williams was not alone in altering the meaning of the Indian documents since he also used the translations of other Europeans as the basis for his dictionary. For example, one of the earlier translators of the *Rig Veda* was Max Müller (1869) who was quoted as using his translation in order to take away the Indians faith in the *Vedas*.[185]

---

[184] See List of Sources.

[185] *Sarasvati & Vidyalankar* (1977) Vol. I, Ch. 4-5

One of the objects of the English translators was to make their translations sound like the original in terms of rhythm and rhyme. This caused Max Müller to observe, "…very often a metrical translation is an excuse only for an inaccurate translation."[186] Ralph Griffith made what sounds like an excuse for poor translations when he noted that *Book Nine*, which is about *soma* and inner powers, was of "intolerable monotony."[187] Perhaps this statement reflects the loss of the original *Techne* more than any other statement could.

There seemed to be an agreement which continues today among the English scholars of Sanskrit that *soma* had to be produced from some unknown and no longer available plant. The Monier-Williams *Sanskrit-to-English Dictionary* defines *soma* as: "the juice of the *Soma* plant, a drug of supposed magical properties." Similarly, the words which referred to the inner production of *soma* are redefined to be pieces of external production equipment.

English translations of the Indian *Haṭhapradīpikā*,[188] (or *Haṭhayogapradīpikā*), a basic handbook for many modern yoga students, contain translated verses that can actually cause physical harm to anyone attempting to follow them. *Book III*: 31-36 (included on the next page) is translated to instruct the student of *hatha yoga* to slice the frenum of the tongue every day until the tongue is free to reach into the sinuses. This translation is so far from the literal meaning that we suspect some collusion, certainly with other Europeans, but also perhaps with some Indians who perhaps wanted to shame the British.

---

[186] Ibid., Vol. I, Ch. 4

[187] Ibid.

[188] Widely used as a reference for modern yoga classes.

The translation is actually so bad that support for the key English words cannot be found in the original text and the reader is left with no other choice than to perceive the words as having been inserted in the text with many of the original terms ignored. (Fortunately, we know that few *yoga* students read beyond the first two books and know of no one who actually tried to get the tongue to reach into the sinuses.)

Indian Literature
The *Rig Veda* (*Ṛgveda*)

The *Rig Veda* must be viewed as a technical description of and instructions for the perfection of the body and mind that were later elaborated upon by the *Tantras* and some of the early *Yoga* writings. The *Rig Veda* is primarily devoted to the inner powers of an evolving individual and since these powers and means for stimulating them are quite varied, the verses describing them are likewise quite varied.

The *Rig Veda* must be read in the same fashion as the early Greek *Techne* writings where the *Rig Veda* consists of a series of convoluted allegories using different terms for the same metaphysical properties or energies. The best example is *soma* or the inner vital energy which is variously called: *amrita*, moon, night, various oblations or fluids, a goddess, virility, a curative drug, wine of immortality, the procreative bull, etc.

Similarly, the inner activating power that utilizes the *soma* or basic energy to manifest some goal or vision is called by several names such as: *Indra*, *Deva*, *Agni*, *Rudra*, *Varuna*, etc. There are two basic methods of generating *soma*, called churning and winnowing, which involve the pelvic area muscles and the lower abdominal muscles respectively. The muscle movement is then stimulated with images and direct physical stimulation of the outer body.

The first reference to the preparation of *soma* is found in *Book One*, *Chapter Twenty-eight* of the *Rig Veda*. We list first our translation followed by the popular translation given by Griffith in 1889. In order to justify the final translation to the reader, we list the original Sanskrit words in order, followed by a colon and a dictionary meaning or another word with the same root or equivalent spelling and its definition. When necessary, we also include a brief commentary.

Book I - Hymn XXVIII - *Indra, etc.*

Book 1:28

1. Wherein, the firm broad floor (abdominal) can become the perfect hot threshing floor with its power of *Indra*, rising up from below from the covered bulb (*linga* or *kanda*)

> *1. There where the broad-based stone is raised on high to press the juices out, O Indra, drink with eager thirst the droppings which the mortar sheds.*[189]

yatra: wherein. ***grāvā***: firm. ***pṛthubudhna***: broad lowest part. ***ūrdhvo***: rising. ***bhavati***: becoming. ***sotave***: bring forth.	***ulukhalasutānām***: ***ulukhala***: ***ulu***: ***ul:*** burning. ***khala***: granary. ***sutanām***: excellently best ***aved***: going down. ***indra***: inner creative god. ***jal***: covered. ***gula***: *linga* or bulb.

Comment: It is difficult for most modern urban readers to identify the broad threshing floor used in the *Rig Veda* as an allegory of their lower abdomen. The *Rig Veda* describes a central post to which the oxen were tied as they were led around and around to step on the grain and separate the husks from the grain. Once the husks and grain were separated, then the process of purification by winnowing began which required picking up the mixture with a scoop on one end of a winnowing basket and then throwing the

[189] Griffith (1896) Book I, Hymn XXVIII

mixture upward into a crosswind, which then blew the chaff away. It is perhaps only after mastering some of the beginning practices of *Yoga* that the full implications of the central post, the threshing by the oxen, the separation of chaff and grain and the final purification with the wind can be fully appreciated. The rising power is obviously a direct reference to the body rather than to a granary.

2. Wherein just so, these two, make and press out (*soma*) from the loins.

> *2. Where, like broad hips, to hold the juice, the platters of the press are laid, ...*

yatra: wherein. ***dvāv***: two. ***iva***: just so. ***jaghanā***: pudenda or hips.	***adhiś avaṇām***: press out. ***kṛtā***: make.

Comment: The two are, of course, the floor of the abdomen and the lower covered bulb (swollen bulbospongiosus muscle).

3. Wherein, with a motion like a woman's motion (of the hips), away and toward, everything is gained.

> *3. There where the woman marks and learns the pestle's constant rise and fall, ...*

yatra: wherein. ***nary***: of a woman. ***apacyavam***: moving away. ***upacyavaṃ***: moving toward.	***ca***: and. ***śikṣate***: learning to do anything.

4. Wherein, restrained churning is controlled in the same manner by pressing to and fro.

> *4. Where, as with reins to guide a horse, they bind the churning staff with cords, ...*

yatra: wherein. ***mantham̩***: churning. ***vibadhnate***: pressing to and fro.	***raṣmīn***: control. ***yamitava***: restrained. ***iva***: same manner.

5. Indeed, the asking for wisdom is connected to the resident giver of wisdom located within the burning inner granary within the body. In this place is manifested a pervading ecstasy like a victorious drum.

> *5. If of a truth in every house, O Mortar, thou art set for work, Here give thou forth thy clearest sound, loud as the drum of conquerors.*

yac: to ask. ***cid***: indeed. ***dhi***: wisdom. ***tvam̩***: skin. ***gṛhegṛha***: servant in a house. ***ulukhalaka***: burning inner storehouse. ***yujase***: connected.	***iha***: in this place. ***dyu***: manifest. ***mattā***: ecstasy. ***vada***: speaking. ***jayatām***: victorious. ***iva***: like. ***dundubhīḥ***: drum.

6. And foremost always longing for and increasing audible breath to and fro. Now to find the protected powers of *Indra*, go to the inner storehouse and produce *soma*.

> *6. O sovran of the forest, as the wind blows soft in front of thee, Mortar, for Indra press thou forth the Soma juice that he may drink.*

utā: and, also.***sma***: always. ***te***: they. ***vanas****: longing.* ***pat***: control. ***vata***: audible. ***vi***: to and fro. ***vāty***: wind of the body. ***agram***: foremost.	***it***: going towards. ***atho***: now. ***indrāya***: powers of *Indra.* ***patave***: protected. ***sunu***: ***su***: to go.***nu*** at once. ***somam***: producing *soma.* ***ulukhala***: hot granary, inner source.

Book IX - Hymn I - *Soma Pavamāna*

Book 9:1

1. Self-directed exhilaration, purifying one's own contained *soma*, generated to be like *Indra* (the god) and protected.

> *1. In sweetest and most gladdening stream flow pure, O Soma, on thy way, Pressed out for Indra for his drink.*[190]

svadiṣ ṭayā: self directed. ***yā***: going. ***madiṣṭha***: very intoxicating or exhilarating. ***yā***: going. ***pava***: purification. ***sva***: one's own.	***soma***: ***dhārayā***: containing. ***indrāya***: to be like *Indra.* ***pātave***: protected. ***sutāḥ***: brought forth, generated.

3. Granting relief without darkness, bringing forth abundant victory, to be successful with friends and wealth.

> *3. Be thou best Vṛtra-slayer, best granter of bliss, most liberal: Promote our wealthy princes' gifts.*

---

[190] Griffith (1896) Book IX, Hymn I

varivodhā: granting relief. ***atamo***: without darkness. ***bhava***: coming into existence. ***maṇhiṣṭo***: exceedingly abundant. ***vṛtrahantamaḥ***: bestowing abundant victory. ***tama***: darkness.	***parṣi***: an assembly, companion. ***rādho***: to succeed. ***magha***: wealth, power.

7. Ten subtle grasping, struggling young maidens (ten fingers) churn to produce the immediate desire.

> *7. Ten sister maids of slender form seize him within the press and hold Him firmly on the final day.*

tam: desire. ***īṃ***: now. ***aṇvi***: ***aṇvī***: the subtle. ***samarya***: struggle. ***ā***: of course. ***gṛbh***: grasping.	***ṇanti***: to churn up. ***yoṣaṇo***: young maidens. ***daśa***: ten.

Comment: Describing the fingers as ten young maidens instructs the reader in how the fingers are to be used. Maidens do not symbolize sex but do symbolized gentleness and patience. The fingers as described later do not do the churning, but rather the stimulation that causes the deep inner churning.

8. The ten fingers stimulate the vessels of the skin to produce a thunderbolt, producing the three powers of the invincible sweet liquid (*soma*).

> *8. The virgins send him forth: they blow the skin musician-like and fuse The triple foe-repelling meath.*

tam: desire. ***īṃ***: now. ***hinvanty***: inciter. ***agruvo***: ***agruvas***: of the ten fingers. ***dhamanti***: ***dhamani***: tabular vessels. ***bākuraṃ***: thunderbolt.	***dṛtim***: skin. ***tridhatu***: three parts giving. ***vāraṇam***: invincible. ***madhu***: sweet liquid (*soma*).

Book IX - Hymn III - *Soma Pavamāna*

Book 9:3

10. The winnowing basket purification has many functions (such as) extracting and generating higher knowledge and power.

> *10. This Lord of many Holy Laws, even at his birth engendering strength, Effused, flows onward in a stream.*[191]

eṣa: seeking. ***sya***: winnowing basket. ***puruvrato***: having many functions. ***ja***: produced by caused. ***jñāno***: higher knowledge. ***janayann***: generating. ***iṣaḥ***: possessing and powerful.	***dhārayā***: possessing and preserving. ***pavate***: ***pava***: purification, winnowing grain. ***sutaḥ***: ***suta***: extracted, brought forth, generated.

Comment: The allegory of the winnowing basket is used to purify the grain placed on the granary floor as in the allegory of 1:28:1

[191] Griffith (1896) Book IX Hymn III

above. The winnowing process follows the pressing or stomping (or churning of the lower abdomen) and separates the grain from the mixture of chaff and grain with moving air. The inner winnowing process requires deep exhalations and therefore requires the addition of the upper abdominal muscles to the lower churning in the abdomen.

Book IX - Hymn XCIII - *Soma Pavamāna*

Book 9:93

1. Together, the ten (fingers) snare, prepare and control the inner power moving quickly like a liquid (but not a liquid) into a container. Produced by traversing and bringing forth from one's chest to attain the extraordinary rewards like a cloud pouring forth its contents.

> *1. Ten sisters, pouring out the rain together, swift-moving thinkers of the sage, adorn him, Hither hath run the gold-hued Child of Sūrya and reached the vat like a fleet vigorous courser.*[192]

sāka: together. ***mukṣo***: ***mukṣija***: net. ***marjayanta***: ***marjya***: to be prepared. ***yantur***: regulator. ***svasāro***: ***sva***: one's own. ***sār***: inner strength. ***daśa***: ten. ***dhirasya dhi***: receptacle. ***rasya***: like a juice.	***adravaj***: ***adrava***: not liquid. ***jāḥ***: produced. ***sūrasya***: ***su***: to bring forth. ***urasya***: coming from the chest. ***droṇaṃ***: as a cloud pouring liquid, a *soma* vessel.

[192] Griffith (1896) Book IX Hymn XCII from Book IX

dhītayo: **dhī**: contain.
tay: to go towards.
dhanutrīḥ: **dhanutṛ**: moving quickly.
hariḥ: carrying.
pary: encircle

nanakṣe: **na**: **nakṣ**: to attain.
atyo: **ati**: extraordinary.
na: **na…na** = reinforcement of statement.
vāji: reward.

Comment: This verse introduces the stimulation of the fingers obviously on the nipples, although unstated.

2. Together the fingers cause the residence of the source to swell separating and producing a flow very difficult to be restrained. The goal of entering the light is connected with the fingers extracting from the chest together with the uniting sensuality which covers the self.

> *2. Even as a youngling crying to his mothers, the bounteous Steer hath flowed along to waters. As youth to damsel, so with milk he hastens on to the chose meeting-place, the beaker.*

sam: together.
mārbhir: **mātṛ**: mother or source.
bhir: **bhi**: to separate.
ṣiṣur: **ṣi**: to cause.
śū : **śvi**: to swell.
vāvaśāno: **va**: like.
avaśāno: residence.
vṛśā: **vṛś**:the fingers.
da: producing.
dhanve: to flow.
puruvāro: **puru**: much.
vāra: difficult to be

yan: **yād**: to be closely united or connected with.
saṃ: together.
gachate: **ga**: being.
maryo: **maryā**: goal.
yoṣām: **yoṣan**: the fingers.
abhi: to, towards.
niṣkṛtaṃ: to extract.
chat: **cat**: to hide one's self, to cover.
kalaśa: breasts, cloud.
usriyābhiḥ:

restrained.	***usriya***: light.
adbhiḥ: ***ad***: consume.	***abhi***: to enter.
bhiḥ: ***bhi***: to separate.	

Comment: The source that swells must be the *kanda* of the *yoni* or the middle of the perineum. The flowing mentioned here and in preceding verses relates to the strong sense of something flowing downward during stimulation as well as upward when the *soma* is demanded or used during some need.

3. Also, filling and swelling the lower abdomen with *soma* like a cloud toward bestowing the nourishing cover.

> *3. Yea, swollen is the udder of the milch-cow: thither in streams goes very sapient Indu.*

uta: also.	***abhi***: to, towards.
pra: before.	***sacate***: ***sa***: bestowing.
pipya: to swell.	***cat***: to cover one's self.
ūdhar: ***udar***: belly, womb.	***sumedhāḥ***: nourishing.
aghnyāyā: being like a cloud.	
indur: ***indu***: *soma*.	
dhārabhiḥ: ***dhāra***: containing.	

Comment: The allegory of a cloud also relates to the vagueness of the flow of *soma*. The flow is felt to come from the region of the *yoni* although the flow itself is more like a gentle rain throughout the body.

Indian Literature
The *Haṭhayogapradīpikā*

The *Haṭhayogapradīpikā* is a book containing a collection with commentaries of older texts written somewhere around the fifteenth century in India. This book is one of the chief references now used by *hatha yoga* groups for their sitting postures, *asana*,[193] and three basic controlled breathing exercises, *pranayama*.[194]

The English translations of the *Haṭhayogapradīpikā* are many times quite different from the original Sanskrit, but this is because the translators made the text reflect modern politically and socially correct views. The word *Haṭha*, for instance, is no longer interpreted as being something violent or forceful, but rather something gentle and relaxing. The word *Yoga*, rather than meaning a metaphysical union, has become synonymous with a prescribed discipline to be followed. *Pradīpikā* meant the striving after illumination or understanding but now is generally assumed to mean the striving for youth and beauty.

Because of the loss of much of the original meanings, some verses were not translated at all, but rather reconstructed around a dominant word. As an example, which has no doubt been a true obstacle to many people, is the verse that advocates slicing the bottom of the tongue so that it can be pushed up into the sinus cavity like a cork to stop sinus drainage. (For clarification, we will present the critical verse in the following text.)

The physiology of the *Haṭhayogapradīpikā* can be illustrated with the simple model of the intense crying and sobbing of a

---

[193] *āsana*: 'sitting'

[194] *praṇyāma*: *pūraka, recaka, kumbhaka* (breathing: inhaling, exhaling, holding the breath.)

doubled-up child mentioned before in Chapter Sixteen. It can also be related to the change in the body during the preparation to overcome some very threatening force or even the contortions and labored breathing experienced during a nightmare or some illnesses.

The first two chapters can be considered to teach methods of developing a supple body capable of finding and maintaining the extreme physical demands put upon the body from deep sobbing. These first chapters also describe techniques for developing the lungs and the controlling muscles. Developed muscles are later required to stimulate the lower abdominal region as well as assisting in the flow of fluids through the body.

The third chapter relates primarily to finding the stimulation and controls found in the tight bending or fetal pose with added stimulation from the excitation of the perineum. The remaining chapters discuss the changes found in the body after the body has been stimulated and include such things as the ringing in the ears heard after stress and how it should be used to increase the inner powers.

The intent of Chapter III is described with verse 8 which states that the goal is to remove the secret inner feminine power from the winnowing basket that contains it. The winnowing basket is the same as the one described in the *Vedas* (*sya*, *śūrpa*) as well as the publicly displayed Liknon basket of the early Dionysians. This basket can be briefly described as being the allegory for the portion of the body commonly referred to as the 'bread-basket.' The next verse initiates the 'great pose' or *mahamudra*. This pose describes how the *yoni* or *kanda* is stimulated by sitting on the foot of one leg while the other leg is extended forward. (In the modern world, the pressure of the foot can be duplicated with sitting on a small soft rolled cloth and bending forward.)

A brief description of the *yoni* and *kanda* is needed, although the reader should consult other books for more information.[195] These words are interesting in that the first dictionary definition of *yoni* is of a female vulva or womb, and the meaning of *kanda* is a bulb or an affection of the female *yoni*. Both words are, however, of masculine gender and generally only applied to the male (although both sexes have the same experiences).

Most modern translators, however, substitute the word perineum for *yoni* and ignore the word *kanda* or the swelling that takes place in the *yoni*. The verses in Chapter III also describe the *yoni* as being a covered cavity containing a moving 'tongue.' These verses describe a control of the source of the moving, very pleasurable perineal feelings encountered in childhood. It is these types of feelings that encourage children to stimulate the perineum or *yoni* with childhood games that involve falling or bouncing on the buttocks as well as the tightening of the perineal muscles.

The tightening or pulsation of the perineal muscles is normally considered to be an involuntary response[196] in adults; however, the *Haṭhayogapradīpikā* teaches how an individual can rediscover ecstatic, powerful inner feelings with a learned control of these muscles. Their development and control leads to the ability to increase pleasure by tightening and loosening certain inner abdominal muscles (i.e., winnowing and churning). During this time, the whole body and mind is stimulated such as would be found during REM dreaming and gradually even beyond that to the experiencing of the higher powers such as found during trauma or emergencies.

---

[195] See Peck (2004) and Peck (1998).

[196] As in coughing, laughing and crying

On the remaining pages of this chapter are a few of the verses from Chapter III that illustrate the basic teachings of the *Haṭhayogapradīpikā* as well as some verses illustrating the original language used to describe these teachings. We will use the same format as we did with our translations of the *Rig Veda* verses. Our literal translation will be given first, followed by a rendition of the popular translations in italicized print. The Sanskrit words (which might be the separated single words of a composite) are then given in order with their roots or origins (when necessary) and the applicable dictionary meanings. When necessary, we also include a brief commentary.

The *Haṭhayogapradīpikā*
Chapter Three - A description of *Mudrās*

9. Persevere and endeavor to obtain the valuable secret concealed in the wicker basket. It is taken out of the winnowing basket (Liknon) truly like a proper pleasurable and uniting feminine presence, as will be explained.

> *9. It needs to be kept secret like a casket of gems and should not be discussed with anyone as a proper woman does not discuss her sexual pleasures.*

gopanīyaṃ: secret.
prayatnena: persevering effort.
yathā: in such a manner as follows.
ratna: anything valuable.
karaṇḍ: a basket of bamboo wickerwork.
akam: to endeavour to obtain.
kasya: *Liknon*. From ***ka***:
kad: to remove the chaff or husk of grain.
sya: winnowing basket.
chinna: taken away or out of.
iva: enclitic such as truly.
vaktavyṃ: to name something.
kulastrī: a good woman.
surata: playful, amorous, coition.
yathā: in such a manner as follows.

Comment: Both the winnowing basket and the granary are used as allegories for describing the production and storage of creative energy. The winnowing basket is an excellent allegory because it illustrates the type of muscle motion used to stimulate the inner powers. Similarly, the abdominal floor compares with the granary floor which resists the downward pressure or force of production. Also, the abdomen stores the energy like a granary stores the grain.

10. The *soma* is pressed out of the *yoni* with the foot together while striving for pleasure. To enter the creative within the self, expand, and hold a pleasurable and fixed stretched out position on the foot.

> *10. Pressing the perineum with the left heel and stretching out the right leg and grabbing the toes.*

pādamula: ***pāda***: the foot. ***mula***: root.	***padam kṛtvā***: to set foot in or on, to enter.
vāma: striving after.	***kara***: creator.
yoniṃ: vulva, receptacle, center of procreative energy.	***bhya***: within the self.
saṃ: together.	***dhāraya***: holding.
pīḍya: ***pīḍ***: to be squeezed or pressed out.	***dṛḍha***: strong.
dakṣinam: that which is right, pleasure.	
prasāritaṃ: stretched out, expanded.	

Comment: This corresponds with experiences found during intense grief or trauma as one gives up control. The grief overwhelms the brain as the body folds into the fetal pose and the breathing becomes extremely labored as one sobs.

The *Haṭhayogapradīpikā*
Chapter Three – *Mahāmudrā*

11. Placing, uniting, holding (and) joining together the narrowest part or neck (of the *yoni*) near the front. In such a manner (as to) force the churning rod to move like a snake, without the churning stick stimulating sexual power.

> *11. Then contract the throat and keep raising the inner wind. As a result of this mudra the kundalini is made to straighten just as a snake does when it is beaten.*

kaṇṭhe: neck, narrowest part.	***vat***: as or like.
bandham: uniting.	***yathā***: in such a manner.
samāropya: placing in or upon.	***daṇḍa***: churning stick.
dhārayed: holding, possessing.	***hataḥ***: forced.
vayu: joining together.	***sarpo***: ***sarpa***: to glide, snake.
mūrdhvataḥ: ***mūrdha***: the front.	***daṇḍā***: a rod, churning stick.
	akāraḥ: without making.
	prajāyate: generative power.

Comment: The churning rod is an allegory for the rising and falling sensations felt in the lower abdomen. The desired feelings are not like a sudden spasm, but rather a very pleasurable shift in rising and falling deep inner muscle tensions. The sexual orgasm is a small example of the feelings. The 'neck' of the *yoni* can only be found as the perineal muscles of men and women become developed. The opening of the (allegorical) neck (not the vagina or anus) can be felt with the effort to pull the perineum up into the lower abdomen. Other verses describe exercises to imagine that some forces or fine substance can be pulled up through that neck which assists in sensitizing the area. This verse is simply stating that the spot felt to open and be moved is the spot to be stimulated. It is not to be confused with the spot (clitoris or penis) that stimulates the sexual response.

12. In that manner the existing female energy of the *kundalini* is forcibly obtained, aroused and made alive. The conquering union of being folded double (bent over) then gives numbness to the lower belly.

> *12. The kundalini shakti becomes straightened and the two nadis become lifeless.*

ṛj: to obtain, acquire.	***tadā***: then.
vi: to set in motion, excite.	***sā***: giving.
bhūtā: being alive.	***maraṇā***: ceasing.
tathā: in that manner.	***vasti***: the lower belly.
śaktiḥ: the female creative power.	***jāyate***: ***jaya***: conquering.
kuṇḍalī: serpent.	***dvipuṭā***: folded double.
sahasā: forcibly.	***āśrayā***: uniting.
bhavet: being.	

Comment: One of the interesting phenomena associated with ecstasy is the deadening of certain areas of the body such as the reduction to pain found as one nears a sexual orgasm. This verse is reassuring the reader that despite increasing effort to squeeze everything together, pain will not result, but only pleasure. The *kundalini* is an allegory of the power of *soma* in the form of a serpent which is credited with deadening parts of the body while also quickening areas or organs of power.

13. Slowly, gently, exhale and strengthen the diffused stream (*kundalini*) to certainly reach the great joy and powers obtained from the granary floor as has been taught.

> *13. Then exhale slowly and not quickly. This is indeed the Mahamudra of the great teachers.*

tataḥ: diffused. ***śanaiḥ***: ***śana***: slowly. ***śanaireva***: ***śanais***: quietly. ***recayenna***: ***reca***: exhalation. ***yen***: by means of. ***tu***: to make strong. ***vegetaḥ***:***vega***: a stream, flood, current. ***iyaṃ***: to approach.	***khalu***: ***khala***: granary floor. ***mahā***: great. ***mudra***: joyous, glad. ***mahā***: great. ***siddha***: perfected, endowed with supernatural faculties. ***pradaśitā***: taught.

Comment: It is assumed the word *khalu* (certainly) is an alteration of the word *khala* which means a granary floor (or as an allegory for the abdominal floor). This would agree with the usage of the word *karand* and *kasya* in verses 9 and 49, meaning wicker basket and winnowing basket (used for extracting and purifying grain). What was 'taught' is no doubt referring to the *Rig Veda* in such verses as 1:28 already quoted.

The *Haṭhayogapradīpikā* Chapter Three – *Khecarī*

32. Within the covered cavity there is a reversal that takes place in which the tongue of fire or *Agni* can be initiated. The motion of the tongue in the cavity obtained by the *mudra* is secured in the middle between the opening and interior.

> *32. To accomplish the khechari mudra, reverse the tongue and thrust it up the back of the throat and turn the eyes toward the eyebrows.*

kapāla: shell, cover. ***kuhara***: cavity. ***jihvā***: tongue (of flame). ***praviṣṭā***: initiated.	***antargat***: going between the entrance and interior. ***dṛṣṭir***: bold, obtained,

viparītā: reversed. ***gā***: undergo. ***bhru***: calling, being.	secured. ***mudrā***: posture. ***bhavati***: state of being. ***khecarī***: ***kha***: hollow. ***cari***: motion.

Comment: The covered cavity is the *yoni* which is felt as an inner softened area in the perineum after control of the perineal muscles is obtained. As the muscles are developed and their sensitivity and responses increased, it is found that the complex muscle structure of the perineum is able to facilitate a moving protuberance (*kanda*) that in its response to stimulation reminds one of the movements of the tongue in children that responds to feelings. The limitation of the motion of the lower tongue becomes quite apparent as one desires to move it further and further outward.

33. (The tongue) is a small tremulous and yielding split-off part that causes a change to begin. Giving higher powers of the cavity when its flames are able to be conveyed as far as the eyebrows.

> *33. This requires lengthening the tongue, by cutting the frenum of the tongue, moving and pulling it until it can reach the eyebrows.*

ched: splitting off. ***cālana***: shaking moving. ***doha***: yielding, offering. ***kalā***: small part. ***kram***: step, approach, stretch over. ***vardayiter***: ***vardhayitr***: one who causes to increase.	***sa***: giving. ***yāvad***: as far as. ***bhrumadhya***: between eyebrows. ***spṛś***: to touch, convey to. ***tadā***: then. ***khecarī***: cavity, path. ***siddhiḥ***: power.

Comment: The tongue, because of its stimulating qualities, was named the tongue of fire or the tongue of *Agni* (the god of fire). One of the evidences of its activity is the rising sense of heat and pressure that ultimately reaches the area behind the eyebrows. At first this pressure raises fears, but later as the pressure feels comforting, one is able to increase it freely.

34. The fluid that is pressed out is pure, pungent and slippery, emitted from a channel resembling an instrument. It is not given to everyone to collect sufficient quantity to fill the cavity with its perfume.

> *34. Use a sharp, lubricated knife in the shape of a cactus plant, cut the frenum a hair's breadth.*

snu: emit fluid.	***sama***: every, some.
hī: to stimulate.	***adāya***: not giving.
patra: ***pātra***: channel of flow.	***tata***: extended.
nibha: resembling.	***stena***: perfume.
śastraṃ: an instrument.	***roma***: cavity.
su: to press out.	***matra***: quantity.
tīkṣṇaṃ: pungent.	***samucchi***: to collect together.
snigdha: slippery.	
nirmala: pure.	

Comment: This is the verse that is generally translated as giving directions for starting the cutting of the frenum of the tongue a hair's breadth every day until the tongue can extend up into the sinuses. This erroneous translation may have arisen because the instrument, *śastraṃ*, does have another definition as a cutting weapon, which other translators may have perceived as of central importance. Also, the description of the tongue of *Agni* can be interpreted as being an inner instrument as well. The authors have experienced that many of the secretions of the body become pleasurably scented during and following the stimulation of the *yoni*. They also noticed a defining characteristic that although it is initially slippery it is not like a lubricant and is very rapidly absorbed into the skin.

49. The penetration of the tongue of *Agni* removes liquid *soma* from the granary floor. The flowing moon fluid (*soma*) is contained within the winnowing basket, home of the vibrant carrier of the sun (*Aruna*).

49. When the tongue enters the cavity, heat is produced.

jihva: the tongue of *agni*.
praveśas: penetration.
saṃbhūta: become, being, existing.
vahni: *soma* (as "the flowing or streaming one").
not: ***nod***: removing.
padita: ***pa***: drinking. ***dita***: pulled from.
khalu: ***khala***: granary floor.

candrāt: moon.
sravati: flowing.
yaḥ: swift, water.
sāraḥ: anything fluid.
sā: giving. ***sa***: junction.
syā: ***sya***: winnowing basket, Liknon.
dama: constrained, home.
rava: vibration, noise.
aruṇa: ***Aruṇa***: The Dawn, the charioteer of the Sun, or the Sun itself.

Comment: We chose this verse since it uses the terms found elsewhere in the text, in the *Rig Veda*, and in the descriptions of Dionysian and *Tantrik* icons. Further, the concept of *Aruna* is also found in the Roman religion of *Sol Invictus* in which the sun, as an allegorical and anthropomorphic creator, was carried across the sky by an intermediary god, thus transferring the heavenly power to its manifesting on earth or in individuals.

Indian Literature
The *Rudrayāmala* or *Parātriṃśikā*

The *Parātrīśikā-Vivaraṇa* was translated, with comments, into English by an Indian scholar, Jaideva Singh, in 1988. This was a translation of the chief writing of the well-known medieval Indian Kashmir Śaivist sage, Abhinavagupta. Abhinavagupta was primarily concerned with *Tantrik* or *Trika* (Trinity) writings and wrote the *Parātrīśikā-Vivaraṇa* to expound upon a very short and much older document known originally as the *Rudrayāmala*. The *Rudrayāmala*, now generally known as the *Parātriṃśikā*, is highly suppressed in India (the reason will become obvious) and we know of no other source for it other than the *Parātrīśikā-Vivaraṇa*.

Our experience with the book is that Abhinavagupta did not translate the ancient and obscure Sanskrit of the *Parātriṃśikā* into more understandable Sanskrit. Instead, Abhinavagupta offered assistance in how it could be translated from its original Sanskrit for comprehension. The *Parātriṃśikā* is obviously written in a secret language to anyone who avoids or is unaware of the metaphysical.

We consider the *Parātriṃśikā* to be a primer defining fundamental terms as well as the basic philosophy associated with most of the ancient world's religions including the existence and nature of the heart in the lower abdomen and of the indwelling goddess or *Aphrodite*. It describes how the nipples cause the secretion of an inner fluid that is no doubt the equivalent of oxytocin that stimulates the *yoni*. The *yoni* is then responsible for the release of *soma*, *amrita* or *ambrosia* that is then controlled by a higher liberating power similar to *Eros* that is described in the *Haṭha-pradīpikā*.

The ancient *Rudrayāmala* was highly advocated by Abhinavagupta, yet he buried it within his medieval *Parātrīśikā Vivaraṇa*,[197] also written in Sanskrit. The text was supposedly an explanation of the *Rudrayāmala*, but our conclusion is that it was really written to explain the *Techne* of the *Rudrayāmala*, which we fully utilized. Abhinavagupta's book is an excellent example of how Sanskrit can be interpreted in many different ways and provides the means for deriving these interpretations.

We were amazed at how well we could use Techne in deciphering old Indian documents that could be used to select and then manifest the inner teachings. These teachings could then be fully supported by science, practical experience, the covering philosophy as well as intuition as required by the early Greeks for judging Techne.

We include eight verses of the *Rudrayāmala*, now generally known as the *Parātriṃśikā*, that clearly evidence methods of inner stimulation of inner churning. The remaining verses can be found in our earlier book;[198] however, their teachings have been well integrated within the preceding text. With the exception of verse nineteen, we have updated our translations of these verses.

A few introductory remarks might be useful. The central subject of the *Parātriṃśikā* or the *Rudrayāmala* is *mātrena*. The reader should have little trouble understanding *mātrena* since it is nearly identical with the concept of the uniting, creative *Eros*. Reading the *Parātriṃśikā* also requires a knowledge of *mudra*, which is the role that can be taken on to change the self, much as a child can become a pirate, lion tamer, or schoolteacher in play.

---

197 *Parātrīśikā*: 'the supreme goddess of the Trinity,' *Vivaraṇa*: 'exposition upon'

198 See Peck (1998).

Reading this text also requires the knowledge of *mantra*, which is the same as Aristotle's entelechy and can be compared with the goal that is being sought for in Proverbs 16:9. *Mantra*, therefore, defines the *mudra* that will be attained. *Matrena* is also referred to as the union of the sun and moon (union of masculine and feminine principles) or the union of *mantra* and *mudra* to effectuate a reality.

The first verse we include below is one that we have published before. It is the clearest description of the heart existing in the lower gut as well as the inner androgyny that exists potentially within both sexes. The organization of the verses is as above, except that we offer no previous translation for comparison; however, commentaries are given to facilitate understanding.

10) The third *Brahma* as a part of the self is obtained by pressing the region of the heart between the loins. Those who do not have the existence as a *yogini* or the state of androgyny, as did the god *Rudra*, cannot break forth.

tṛtīyaṃ: the third. ***brahma***: a god. ***suśroṇi***: ***su***: pressing out ***śroṇi***: hips, loins. ***hṛdayaṃ***: region of the heart. ***bhairavā***: a form of *Shiva*. ***ātmanaḥ***: instrument of the soul, being one's self.	***etan***: this. ***nā***: no, not so. ***yoginī***: female *yogi*. ***jāto***: produced. ***nārudro***: ***nā***: no, not so. ***rudra***: god half male and half female. ***labhate***: obtaining. ***sphuṭam***: appear suddenly.

Comment: The third 'nature' of *Brahma* can be described as being the state of androgyny. Any act of creation must begin with the opening of the body and soul to a vision. This obviously is a feminine act and not a normal masculine act of conquering or subduing. The Greek beautiful and responsive *Aphrodite* as the soul is a perfect model for this verse. The center for this powerful feminine force is universally declared to be the heart; however, *Tantra* places this heart between the thighs, which most women could agree is experienced by tightening the thighs together. The masculine aspect of creation is as described by the inner God *Eros* in Greek philosophy. A *yogini* is normally considered to be a female *yogi* but also one with special mystical powers. The story of the androgynous god *Rudra* was highlighted by how, after being locked deep in a well, he escaped through the creation of his own freedom, which is of course highly suggestive of escaping our own entrapment with life.

There are considerable references to the '*yoni*' or 'female organ' in men within *Yoga* writings and it is associated with the location of the *kanda* and *Shiva Linga*; however, because of this location, and the reluctance of most translators to acknowledge a female center in men, the word *yoni* is often translated as perineum although the standard Sanskrit dictionary does not list this as an alternate.

19) The unseen fertile fluid moves, and thus, with this motion, reality is known, a portion of his powers come into their own existence, he is a *yogī* (feminine), he is also initiated.

adṛṣṭa: unseen.
mandalo: circular. ***pi***: to move.
evaṃ: thus.
yaḥ: ever flowing.
kaś: to go. ***cid***: to divide.
vetti: ***vid***: to be strong.
tattvataḥ: true or real state. ***āta***: going.

sa: giving.
siddhi: accomplishment, fulfillment. ***bhāg***: a part, portion, share. ***bhaven***: becoming, being.
nityaṃ: perpetual.
sa: giving.
yogī: *yogini*: female *yogi*.

	ca: and. ***dīkṣitaḥ***: initiated into.

Comment: The hidden creative fluid is spoken of in many places in the old *Yoga* literature and called by many names. It has been called the flow of the *kundalinī* in more recent literature. The resulting portion is that portion that is required to meet the particular challenges at that time. The early Christians called it the inner Holy Spirit. The "initiation" or *dīksha* corresponds to the taking on of a new life or a dedication.

27) Bring the physical body: the head, sexual organs and the center into the state of bestowed impressed goodness and enchantment effected by either touching and clenching the nipples or caressing the breast.

mūrdhni: ***mūrdhan***: the head, the first part of anything. ***ni***: to bring into any state or condition. ***vaktre***: ***vaktra***: mouth, beginning, the initial or first term of a progression. ***ca***: and. ***hṛdaye***: center or heart. ***guhye***: concealed, sexual organs. ***mūrtau***: material form. ***tathaiva ca***: in like manner.	***nyāsaṃ***: placing, impress. ***kṛt***: effecting: ***vā***: either. ***śikhām***: nipple. ***baddh***: clenched. ***vā***: or. ***sap***: caress. ***ta***: the breast. ***viṃ***: ***vin***: bestow. ***śati***: goodness. ***mantritām***: enchantments.

Comment: This verse may well be the secret stage of *Yoga* that the masters would teach their ready students by allowing them to observe them in their meditation. It certainly would be a far better method of teaching than attempting to describe it. This verse is likewise an example of how the inner meaning of some verses was hidden. The words *sap ta viṃ* and *śati* are compounded together to also produce the word for 'twenty-seven' which is much easier to read than to separate the words. The word *śikhām* has the fundamental meaning of something pointed, which leads to its usage to describe the nipple. It also is used to describe such things as a peacock's crest or comb which probably suggested to some translators that it could be applied to the tying of a lock of hair for a religious symbol. This leads to the translation that a tuft of hair is to be tied with twenty-seven *mantras*. This may be representative of the problems encountered with the *Rudrayāmala*. However, it is only when the Sanskrit is unscrambled properly that a sensible meaning appears using the literal meanings of the words with no externally added words.

28) (That) single region produces close protection and union, giving thee three beginning controls: procuring vibrations (like gnosis), removal of obstacles and reducing passions.

ekaikaṃ: single.
tu: to have power.
diśām: region, area.
bandha: to unite, produce. **daśana**: a tooth, a bite, armor.
am: little.
api: closeness.
yojayet: to unite.
tāla: control.
dattva: giving to thee.
trayam: triple. **purā**: at first.
saśabdaṃ: **sa**: procuring: **śabda**: vibrations.
vighna: removal of obstacles.
śāntaye: reducing passions.

Comment: (No comment needed.)

29) The nipples need to be considered (with) intense yearning toward the three controls to strengthen and prolong *soma* (energetic liquid existing to strengthen) to fulfill going a step either to the whole sign of gender as a limit or to …

Śikh: nipples. ***saṃṃkhyā***: to be considered as. ***abhijap***: intense yearning for. ***tena***: in that direction. ***toy***: liquid, water. ***ina***: energetic. ***ābhu***: to exist. ***ukṣ***: to strengthen.	***tataḥ***: prolonged, protracted. ***puṣ***: to fulfill. ***pādikam***: only one part. ***kramāt***: step, going. ***sarvaṃ***: whole. ***liṇge***: sign, the sign of gender. ***vā***: either. ***sthaṇḍile***: a limit. ***atha***: now. ***vā***: or.

Comment: (No comment needed.)

30) Hide and become playful, procuring a yearning there to gain and endure the performance of sitting. Under those circumstances with protracted sitting, *soma* is created, purified and given.

cat: to hide one's self. ***urd***: ***ūrda***: playful. ***śa***: ***sa***: procuring, bestowing. ***abhijāp***: to pray for. ***tena***: where there. ***puṣ***: to be nourished. ***peṇā***: to endure, embrace. ***āsana***: sitting. ***kalpan***: fashioning or performing.	***tatra***: under those circumstances. ***sṛṣṭiṃ***: creation. ***yaj***: to offer, sacrifice, give. ***vīraḥ***: an intoxicating beverage (*soma*). ***punar***: to purify. ***eva***: just so. ***asanaṃ***: sitting. ***tataḥ***: protracted.

Comment: (No comment needed.)

31) Creation of the world begins with making a bowl shape following a strong dedication to begin. Filled with the complete truth, everything maintains the satisfaction.

sṛṣṭiṃ: creation of world, creation. ***tu***: strong. ***saṃ***: ***sam***: union of. ***puṭī***: ***puṭa***: hollow space, buttocks. ***kṛtya***: make. ***paścād***: following. ***yajanam***: consecrate. ***ārabhet***: begin. ***sarva***: everything, whole. ***tattva***: truth.	***susampūrṇam***: completely filled. ***sarvābharaṇa***: everything. ***bharaṇa***: maintaining. ***bhūśitām***: ***bhūśitum***: to strive after.

Comment: The phrase of making a bowl shape (or *yoni*) can only be done in the perineum. Beginning exercises use the placing of a heel to effectuate this hollow, followed with the development of inner muscles that can pull an equivalent area of the perineum up into the body. It should be noted that *sampuṭa* is also defined as "a kind of coitus." (Obviously this refers to how 'two hollows' or developed *yonis* could join together, which is not the normal mode of sexual intercourse.)

32) Submit to the inner great feminine power by caressing and arousing the breast or perineum to stimulate the goodness (manifesting *mantra*) of visions or *mantra* according to the inner power and guided by the gods as indicated by the expanding sensory perceptions and the many evolutionary feelings.

yajed: consecrate.
devīṃ: a female deity.
maheśānīṃ: ***maheśa***:
great lord or god.
ānīṃ: to bring forth.
saptaviṃśa: ***sap***: caress.
ta: the breast, the womb.
viṃ: ***vin***: incite, arouse, bestow.
śati: goodness.
mantritā: enchantments.
tataḥ: that.
sugandhi: ***suga***: a good path, easy or successful course.
dhi: to hold.
puṣ paistu: ***puṣpa***: expanding.
yathāśaktyā: according to power or ability.
samar: accompanied by the gods.
cayet: ***caya***: a cover.

Comment: (No comment needed.)

Maslow's *Hierarchy of Needs*

(The following *needs* proposed by Maslow are sorted to fit the approach of this book.)

The basic needs for continued life on earth are:

- Air
- Water
- Food
- Shelter
- Sex

To start to develop as a member of a society requires:

- Safety
- Security
- Stimulation
- Orderliness
- Freedom

The process of becoming a social individual requires:

- Self Esteem
- Esteem from others
- Love
- Belongingness

To find one's place in the social world requires:

- Perfection
- Necessity
- Completion
- Simplicity
- Richness
- Justice
- Order
- Playfulness
- Effortlessness
- Self Sufficiency

To find the capability to direct one's own life requires:

- Truth
- Vitality
- Goodness
- Beauty
- Individual Identity

For a statement of how a righteous world filled with union and *Eros* might be described, consider the following taken from the book, *Joy and Evolution: A daring introduction to the technology of Joy and its associated Evolution.*

The Scaffold for Perfection

The outer world reflects the inner.
Both come from mind. Mind is two.

~

One is masculine, the other feminine.
The masculine is the matrix the feminine is the praxis.
Cultivation and its bondage
are the forced juncture of the two.
Freedom starts with
finding the masculine and feminine.

~

The feminine is that which was recast as prolificacy.
The masculine is buried under conceit.
The two must then be carefully
separated and then renewed.

~

The masculine is activated with fervent yearning.
Yearning starts with the loss of immediacy.
The feminine is regenerated with
boundless, childish joy.
With unrestrained sensual stimulation,
the future is desired.

~

Seeking more, masculine and feminine are increased.
The masculine begins the vision of heaven.
The feminine becomes a voluptuary.

~

The two recombine to form the reality of heaven.
Inhabitants of heaven seek oneness.
This begins with union of vision and joy.

~

Eternal evolution and ecstasy is the gift.

Samples of Poetic *Techne*

(Untitled)

Dickinson, E., 1870

We never know how high we are
Till we are asked to rise
And then if we are true to plan
Our statures touch the skies-

The Heroism we recite
Would be a normal thing
*Did not ourselves the Cubits warp**
For fear to be a King-

"That's My Weakness Now"

Excerpt, Green, B., and Stept, S., 1928

Love, love, love, love,
What did you do to me?
The things I never missed
Are things I can't resist.

Love, love, love, love,
Isn't it plain to see
I just had a change of heart;
What can it be?

* Warping cubits has a meaning of changing the standards of measuring something.

List of Sources

Abhinavagupta. (1988) *Parātrīśikā-vivaraṇā.* (J. Singh, Trans.). Delhi, India: Motilal Banarsidass

Angus, S. (1967) *The religious quests of the Graeco-Roman world.* New York, NY: Biblo and Tannen (Original work published 1929)

Apel, W. (1944) *Dictionary of music.* Cambridge, MA: Belknap Press

Aristotle. *Categories.* (E. M. Edghill, Trans.) http://classics.mit.edu/Aristotle/categories.3.3.html

Aristotle. *Metaphysics.* (W. D. Ross, Trans.) http://classics.mit.edu/Aristotle/metaphysics.html

Aristotle. *Nicomachean ethics.* (W. D. Ross, Trans.) http://classics.mit.edu/Aristotle/nicomachaen.html

Aristotle. *On the soul.* (J. A. Smith, Trans.) http://classics.mit.edu/Aristotle/soul.1.i.html

Aristotle. *Politics* (B. Jowett, Trans.) http://classics.mit.edu/Aristotle/politics.html

Cumont, F. (1956) *The mysteries of Mithra.* New York, NY: Dover (Original work published 1903)

Descartes, R. (1911) *Meditations on first philosophy.* (E. S Haldane, Trans.) (Original work published 1641) http://selfpace.uconn.edu/class/percep/DescartesMeditations.pdf

Dickinson, E. (1961) *Final harvest: Emily Dickinson's poems.* Boston, MA: Little Brown & Co. (Original work published c.1870)

Digambaraji. (Ed.) (1998) *Haṭhapradīpikā of Svātmārāma.* (2nd edition). Lonavala, Dist. Pune: Kaivalyadhama, S.M.Y.M. Samiti

Faliero, M. (Ed.) (1998) *Parâtrîmsi'kâ.* In DSO Sanskrit Archive. Originally retrieved from http:/w3.uniroma1it/studiorientali/indologia/gandharvanagaram/00skr) on1/25/2007. Currently available at http:/gretil.unigoettingen.de/gretil1_sanskr/6_sastray/3_phil/saiva/paratriu.htm

Frank J. & Frank, J. (1961) *Persuasion and healing.* Baltimore, MD: Johns Hopkins University Press

Goble, F. (1970) *The third force.* New York, NY: Grossman

Griffith, R. (1889) *Hymns of the Ṛgveda.* New Delhi, India: Munshiram Manoharlal Publishers (Original work published 1889)

Hausman, C. (1993) *Charles S. Peirce's evolutionary philosophy.* Cambridge, England: Cambridge University Press

Herbermann, C. G. (Ed.) (1907) *The Catholic encyclopedia.* New York, NY: Robert Appleton Co.

Hesiod. *Theogony* in *The Homeric hymns and Homerica* (1914) (H. G. Evelyn-White, Trans.). Cambridge, MA: Harvard University Press

Jahn, R. & Dunne, B. (1989) *Margins of reality.* San Diego, CA: Harcourt Brace

Kaviraj, S. (2005) The sudden death of Sanskrit knowledge. *Journal of Indian Philosophy*, 33:119-142

Kegel, A. H. (1951) Physiologic therapy for urinary stress incontinence *JAMA*, *146*(10):915-917

Lapedes, D. (Ed.) (1978) *Dictionary of scientific and technical terms*. New York, NY: McGraw-Hill

Liddell, A.G. (1996) *Greek-English lexicon: Abridged from Liddell & Scott's Greek-English lexicon*. Oxford, UK: Clarendon Press (Originally published 1891)

Lowrie, W. (1923) *Monuments of the early church*. New York, NY: Macmillan

Lu K'uan Yu (Trans.) (1973) *Taoist yoga: Alchemy & immortality*. York Beach, ME: Weiser

Monier-Williams, M. (1990) *Sanskrit-English dictionary*. Oxford, UK: Clarendon Press (Original work published 1899)

Nilsson, M. (1957) *The Dionysiac mysteries of the Hellenistic and Roman age*. Lund, Sweden: Gleerup

Ortega y Gassett, J. (1993) *The revolt of the masses*. New York, NY: W. W. Norton

Parsons, J. D. (1896) *The non-Christian cross*. London, UK: Simpkin, Marshall & Co

Peck, R. & Peck, T. (1988) *The stone of the philosophers*. Windham Center, CT: Personal Development Center

Peck, R. (1998) *The golden triangle*. Lebanon, CT: Personal Development Center

Peck, R. (2001) *Creating heaven on earth*. Lebanon, CT: Personal Development Center

Peck, R. (2001) *Finding power*. Lebanon, CT: Personal Development Center

Peck, R., Cassinari, L. & and Gavlick, C. (2004) *Joy and evolution*. Lebanon, CT: Personal Development Center

Peirce, C. (1878, January) How to make our ideas clear. *Popular Science Monthly*, *12*, 286-302.

Plato. *Cratylus* in *Plato in twelve volumes,* Vol. 12 (1921) (H. N. Fowler, Trans.) Cambridge, MA: Harvard University Press (Original work published c. 385-378 BCE)

Plato. (2013) *Republic*. (C. Emlyn-Jones & W. Preddy, Trans.) Cambridge, MA: Harvard University Press (Original work published c. 385-378 BCE)

Plato. (1925) *Timaeus* in *Plato in twelve volumes*, Vol. 9 (W. R. M. Lamb, Trans.) Cambridge, MA: Harvard University Press (Original work published c. 385-378 BCE)

Plato. (1925) *Symposium* in *Plato in twelve volumes,* Vol. 9. (H. N. Fowler, Trans.) Cambridge, MA: Harvard University Press (Original work published c. 385-378 BCE)

Plotinus (1991) *The enneads*. (S. MacKenna, Trans.) London, UK: Penguin Books

Poe, E. A. (1983) *The unabridged Edgar Allen Poe*. (T. Mossman, Ed.) Philadelphia, PA: Running Press

Prigogine, I. (1996) *The end of certainty*. New York, NY: The Free Press

Putnam, R. (2000) *Bowling alone*: The collapse and revival of American community. New York, NY: Simon and Schuster

Sarasvati, P. & Vidyalankar, S. (Trans.) (1977) *Ṛgveda Samhita*, New Delhi, India: Veda Pratishthana

Ring, K. (1982) *Life at death*. New York, NY: Quill

Robinson, J., (Ed.) (1988) *The Nag Hammadi library in English*. San Francisco, CA: Harper & Row

Schiller, F. (1967) *On the aesthetic education of man*. (E. Wilkinson & L.A. Willoughby, Trans.) Oxford, UK: Clarendon Press (Original work published 1795)

Schrödinger, E. (1992) *What is life?* Cambridge, UK: Cambridge University Press (Original work published 1944)

Stept, S. H. & Green, B. (1928) That's my weakness now. [Recorded by C. Edwards] http://www.heptune.com/lyrics/thatsmyw.html

Ulansey, D. (1989) *The origins of the Mithraic mysteries*. New York, NY: Oxford University Press

Vertosick, F. Jr. (2002) *The genius within*: Discovering the intelligence of every living thing. New York, NY: Harcourt

Westbury-Jones, J. (1939) *Roman and Christian imperialism*. Port Washington, NY: Kennikat Press

Woodroffe, J. (1964) *The serpent power* (7th edition). Madhas, India: Ganesh & Co. (First edition published 1919)

Index

www.ingramcontent.com/pod-product-compliance
Lightning Source LLC
LaVergne TN
LVHW021455080426
835509LV00018B/2287
* 9 7 8 0 9 1 7 8 2 8 1 1 9 *